Understanding Statistics

As A Language

By: Robert D. Andrews, EdD
Larry A. Standridge, PhD

$69.95
ISBN 978-1-61539-808-9
56995>
9 781615 398089

Student Edition 9

CONTENTS

APPENDIX

ILLUSTRATIONS

TABLES

PROCEDURES

Vita: Robert D. Andrews, EdD

Robert D. Andrews, Dean of Graduate and Professional Studies at Bryan College, has taught statistics at the doctoral level for nearly a decade. He has also taught Hermeneutics and Research, an introductory course to the hermeneutics of research reporting and design. Dr. Andrews' educational background includes a BA in history, an MDiv from Trinity Evangelical Divinity School, an MA in Educational Psychology from Tennessee Technological University, and an EdD in Educational Leadership from The University of Tennessee (Knoxville).

As a member of the Graduate Research Council at Oxford Graduate School, Dr. Andrews supervised numerous dissertation research projects. In addition to his dissertation supervision Dr. Andrews co-authored *Understanding Research Communication* with Larry A. Standridge, PhD and Helen Morgan, PhD, *From Questions to Answers: Principles and Methods of Quantitative Research* with Richard P. Walters, PhD, and *Understanding Statistics as a Language* with Dr. Larry A. Standridge.

Dr. Andrews is currently conducting research analysis on two independent projects: 1) A Sports Manifesto that is designed to investigate the relationship between values and attitudes about sports and 2) Faculty Perceptions of Faculty Development Practices in the Counsel for Christian Colleges and Universities—a 20-year longitudinal, follow-up study based on his 1980 dissertation. Dr. Andrews co-authored *Needs Assessment for the Ryan White Title III Planning Grant* for Chattanooga Cares with Dr. Richard P. Walters. Dr. Andrews has also supervised numerous research projects in various businesses in the Chattanooga area. Included among those projects were feasibility studies for ISO 9000 certification, human error management at the TVA Sequoyah Nuclear Plant, and research for an Affirmative Action Plan in a textile plant.

At the undergraduate level, Dr. Andrews has taught Greek, New Testament, Bible, Theology, Interpersonal Communication, Group and Organizational Behavior, Research, Ethics, Creative Problem Solving, Statistics, Principles of Management, and Philosophy of Culture. Dr. Andrews has also taught internationally at Donesque Christian University in the Ukraine, Calvary Chapel Bible College in Hungary, and at Oasis University in Trinidad.

Dr. Andrews lives in Dayton, Tennessee, where he is a founding board member of the YMCA, Secretary of the Rhea Dayton Tennis Association, member of the Glenn Draper Singers, a tennis player, private pilot, and motorcyclist.

Vita: Larry A. Standridge, PhD

Larry A. Standridge has been involved in Adult Education and Educational Research for twenty-five years. After earning his BA in Communications from Baylor University, Dr. Standridge completed an MA at Southwestern Baptist Theological Seminary and his PhD at Ohio State University in Adult Education and Educational Research.

Dr. Standridge has served various institutions as Director or Dean of Continuing Education including John Wesley College, Owosso, MI, Lansing Community College, The University of Tennessee at Martin, and Sterling College, Sterling, KS. As Supervisor of Graduate Research at Oxford Graduate School, Dr. Standridge directed doctoral research for several years. His personal research has focused on the History of the Southern Baptist Denomination and has resulted in several research reports to the Oxford Society of Scholars. Dr. Standridge is also involved in ongoing research for Oxford Graduate School on the nature of dissertations produced by Oxford students. As the person on campus responsible for Institutional Research and Evaluation, Dr. Standridge consistently monitored Institutional Effectiveness for Oxford Graduate School.

Dr. Standridge lives in Martin, Tennessee where he has been active in his local church, real estate development, and telecommunications.

CHAPTER I

STATISTICS AS A LANGUAGE

Communications Tool

Statistics is a tool for simple and effective communication of information, especially new information. Statistics can be compared to using a microscope, telescope, overhead projector, videotape, film, textbook, or a computer. A researcher makes a decision about the best way to present quantitative and qualitative information. One statistical formula may be as effective as twenty pages of text in reporting quantitative information.

Definition: S*tatistics* is an advanced form of communication -- a specialized form of communication using numbers, letters, and symbols as a medium. Although the word statistics may refer to numerical data, it will be used more generally in the course of study to refer to the body of principles and procedures developed for the collection, classification, and summarization of data.[1]

Statistics is a substantial component of research language in all the behavioral sciences. Statistics is involved in the research effort from the beginning to end. At the outset statistical terms play a large part in the definition of problems for research. Specific plans for an investigation are formulated to permit statistical evaluations. Data from the investigation are analyzed by statistical methods. Finally, reports of findings are often expressed in statistical terms. Because of accuracy, versatility, and effectiveness in communication, statistics figures prominently in the language, in which researchers theorize, plan, evaluate, speak, and write.

---

[1] A descriptive statistic is an index number that summarizes or describes a set of data....Inferential statistics is a method that takes chance factors into account when samples are used to reach conclusions about populations. (Spatz, 1997)

Definition: *Data* is a collection of observed values representing one or more characteristics of objects or units. Most social data are in the form of categorical frequencies, cases in defined classes or categories.

Language and Academic Studies

Early in the development of the university, most academic material was written in Latin and later in German and French. Graduate education, especially at the doctoral level, required a working knowledge of Latin, German, or French. Since major academic works were in languages other than English, language study was required. Because of advances in the translation of primary data and because the intellectual center of the world shifted to the United States around 1940, the study of other languages in graduate schools was minimized.

In education and industry, participants encounter statistical information in reading previous research. Further, participants present and interpret their research using statistics in order to understand the findings and to communicate the findings clearly. The use of statistics to report research is not as much a mathematical use as a language use. The course of study is not intended to make participants expert statisticians, rather the study is intended to make participants communicators of statistics, both as receivers and senders.

The language of statistics is a part of the culture of academia. In the academic culture, statistics is the primary means of communicating research data. As in any language, the language of statistics has grown and evolved. Early English has evolved dynamically into present-day English. In the same way, early statistical language changed dynamically. While the symbols of the language remained the same, the language has progressed from manual mathematics to computer assisted analysis.

The dynamic of change is similar to the changes in the culture of the automobile. At one time mechanical knowledge of the car was necessary in order to own and drive a car. The dynamic change in ownership of automobiles has changed all of that. The typical driver of today does not need to know details of the mechanics of automobiles in order to drive. All a driver needs to know is where to put in the gas and how to manipulate the controls.

A pilot in ground school learning to fly is taught to compute fuel usage, navigation, weight and balance data, and airspeed by means of a circular slide rule. When the student pilot steps into an airplane, most data management is preformed by a computer.

Ordinarily statistics had the same dichotomy between what was taught in the classroom and what was expected in actual use of statistics in a research project. Knowing how to compute a standard deviation or Chi square did not help when a researcher was confronted with a totally different problem such as converting questionnaire responses or quality control information into data for computer analysis. A researcher may have a brilliantly designed project that was approved by a committee, and be without sufficient logistical knowledge to bridge from questionnaire or production information to statistical analysis and presentation. The bridge from design to analysis is missing in research literature. The resulting gap has frustrated would-be students of statistics for generations. That gap is perceived as a lack of understanding of the mathematics of statistics, but it is really a gap in the language of statistics. Math is no longer a major issue. Just as an experienced pilot throws away the circular slide rule, the statistician lays aside the calculator, pen, and pencil and uses the computer.

The computer does the math; the researcher needs primarily to know the language of statistics. The language of statistics is not difficult. Statistical language can be broken down into three basic components: type of data, number of groups studied, number of measurements. When the data, groups, and measurements are known, the resulting statistical analysis is largely pre-determined, and the computer can easily work the statistical math. The key is understanding the language of data, groups, and measurements.

Behavior as a Statistic

Behavior can be expressed in statistical values, such as sums, means, proportions, differences, or ratios. The student of behavioral science must think about the behavior of people and animals in terms of statistical values. Understanding of behavior as a statistic must develop, not around isolated acts of an individual, but rather around the mean as an evaluation of a number of acts, whether from an individual or a group, or around the proportion of acts that are observed to be in a certain category or around the differences among individuals and the differences between groups.

Functions of Statistics

Statistical methods in behavioral and industrial sciences serve four related functions. The functions of summarizing, describing, generalizing and experimenting describe the acquisition and communication of new knowledge through research.

Summarizing

The first function of statistics is to summarize research data for communication. A researcher can communicate a mass of information best by first condensing and summarizing the information.

Definition: A *summary* is a condensation in which major points of the original account are retained and from which minor points are excluded. The loss of detailed information is regrettable but necessary if the main features of a large amount of data are communicated.

Summarization may involve arranging a set of numbers into a grouped frequency distribution table as an initial step in developing a graph. Central tendencies can then be easily visualized.

Definition: *Summary statistics* is a tool for collecting, condensing, and organizing numerical data.

Describing

A second function of statistics is description. Statistical descriptions include measures of central tendency, variability, correlation and regression, the normal distribution, standard scores, and graphing methods. Statistics in a text or table communicates with clarity the description of a set of numbers.

Generalizing

The third function of statistics is generalizing beyond the limits of a specific investigation to situations that have not been observed. Achieving the first two functions of summarizing and describing is a somewhat limited goal but one that is essential to the meaningfulness of the third function. If what has been observed cannot be expressed clearly and adequately, attributing to the unobserved certainly cannot be done with clarity and meaningfulness.

Results that remain particular usually garner little interest. Particular results that can be generalized, or more accurately, can be interpreted generally garner a great deal of interest. Arguing from the particular to the general is inductive inference. In the long history of arriving at generalized findings, the best solution available has been statistical inference.

Inferential statistics is a valuable body of theory and procedure for generalizing from particular data. Although inference from the particular to the general is subject to error, statistical inference is the best and the closest to an error-free procedure available. A statistical inference is essentially the act of making and communicating a decision about research results.

Definition: *Inferential Statistics* goes beyond description to draw conclusions or make predictions by interpreting the data. The two classes of inferential procedures are estimation and hypothesis testing. Frequently used inferential statistics are *t* statistics, *F* ratio, and chi-square,

Experimenting

Experimental design involves the manipulation of an independent variable in a study to determine the effect on a dependent variable, which is the variable of interest. The group to which the independent variable is assigned is known as the experimental group; the group in which no manipulation occurs is known as the control group. For example if the Fly-By-Night Pharmaceutical Company wanted to test the XYZ Elixir for stopping the hiccups, researchers would give the XYZ Elixir to one group of hiccuping people and a placebo to another group of hiccuping people. Stopping the hiccups would be the dependent variable, administering XYZ Elixir would be the independent variable, the group receiving the XYZ Elixir would be the experimental group, and the group receiving the placebo would be the control group. Testing for differences between the experimental group and the control group allows conclusions to be drawn about the effectiveness of the Elixir.

By contrast, survey research is retrospective – the effects of the independent variables on dependent variables are recorded after they have occurred. In this instance the Fly-By-Night Pharmaceutical Company might use a survey instrument to determine if people who used the Elixir reported different results in the treatment of their hiccups than people who did not use the Elixir.

Measurements

Numbers can be used in at least three different ways, depending on the level of measurement. When numbers are used in any fashion, data is created. Numbers may have identity, rank, and additivity. Series of numbers can be used (1) to categorize at the nominal level of measurement, (2) to rank or order at the ordinal level of measurement, and (3) to score at the interval level of measurement

Definitions: *Measurement* is the assignment of numbers to observations.

Counting

The number of observations in a category is known as the frequency. A frequency distribution is a tabular display that displays how many individuals possess each value of a particular variable.

Percentages are used to compare groups of unequal size on an equitable basis. The common base for percentage is 100. Regardless of the number being compared in two or more categories, the base of comparison is 100 or more accurately 100% is the whole for each category.

Proportions have a base or total of 1.0. Proportions are always parts of something and can never exceed the total, which is 1.0. Proportions are frequently exchanged for probabilities.

Ratios describe rates and relationships and are fractions. The base, as with a proportion, is 1.0. Proportions are restricted to the relationship of a part to the total; ratios are helpful as index numbers. For example, IQ is an index number of the rate of general mental growth to chronological age.

Nominal Scale

The nominal scale of measurement is the most limited type of measurement and involves the process of naming or labeling. Nominal measurement can be described as classifying observations. All observations placed in the same class are treated as equal. This is the property of identity. The categories must be mutually exclusive; that is, each observation is placed in one and only one category. The categories must also be exhaustive; that is, a place must be provided for each observation. Nominal data cannot be graded, ranked, or scaled for qualities such as better or worse, higher or lower, more or less. Observations in each category can be counted, which provides a frequency. Demographic data such as groups of ages, gender, nationality, or denominational affiliation are nominal scale data.

There are statistical methods designed to analyze nominal or categorical data. Chi-Square is a primary method of analyzing nominal data.

Ordinal Scale

The ordinal scale of measurement is used when observations can be placed in order, based upon a characteristic. While an order can be observed, the exact measurement between two observations cannot be determined. Thus, the distance between A and B may be much greater than the distance between B and C. The ordinal scale allows for the characteristics of more or less of a characteristic, but not how much more or less. An ordinal scale does not have an exact zero nor equal units of measurement. Likert scale items can produce ordinal scale data when the responses are ranked, for example, from one to seven and labels such as 'Agree' and 'Disagree' are attached to each end of the scale. The distance between a response of 3 is only considered to be greater than a response of 2 and smaller than a response of 4. The response of 3 is not equidistant from 2 and 4.

Because the categories of observations have rank, additional statistical methods of analysis are available including correlation and tests for significant difference. In the survey research example given above a Likert scale item might ask the respondent to circle a number from 1 to 7 with 1 indicating little or no relief from the hiccups and 7 indicating total relief. The researcher could then analyze the differences in the responses among various demographic groups.

Interval Scale

The interval measurement scale involves the application of a scale with equal units and an arbitrary zero. With equal units of measurement, most of the useful numerical operations, such as addition, subtraction, multiplication, and division can be performed. Temperature readings on a Celsius or Fahrenheit thermometer, and individual intelligence test scores like the Stanford-Binet or the Wechsler tests are examples of interval data. The measurement of zero on interval scales does not mean the absence of the trait being measured, simply that the scale was not sensitive enough to measure levels of the variable below a certain point. Interval data analysis involves the use of powerful statistical methods such as correlation, t-Tests, and Analysis of Variance (ANOVA).

Table 1 -- Scales of Measurements Summary [2]

Scale of Measurement	Name or Categorize	Rank Order	Discern Equal Differences	Make a Ratio of Two Values
Nominal	X			
Ordinal	X	X		
Interval	X	X	X	
Ratio	X	X	X	X

Ratio Scale

A ratio measurement scale is very much like an interval measurement scale except the observations have an absolute zero. Dollars or cents, yards or feet, minutes or seconds and cash on hand are all ratio measurements or

[2] Table adapted from Elmore et al. (1997)

measurements on a ratio scale. Practically speaking, ratio data are equivalent to interval data and may be analyzed by many of the same statistical methods.

Hypothesis Testing

In the behavioral sciences, research is conducted to determine the acceptability of hypotheses derived from theories of behavior. The researcher develops a hypothesis from a theory and collects empirical data to determine the acceptability of the hypothesis. The empirical data may lead to retaining, revising or rejecting the hypothesis and/or the theory from which the hypothesis was developed. (Siegel 1956)

Definition: A *research hypothesis* is a statement of theory, previous research results, or other information that leads to an anticipated outcome.

Definition: A *null hypothesis* is a hypothesis of no difference. According to the null hypothesis, any observed difference between samples is regarded as a chance occurrence resulting from sampling error alone. If the null hypothesis of no difference is rejected, the alternate hypothesis that a true population or sample difference does exist is accepted.

Definition: An *alternative hypothesis* is a claim about a population parameter that will be true if the null hypothesis is false.

In order to make an objective decision about whether or not a particular hypothesis is confirmed by a set of data, objective procedures are needed. Following are a set of procedures,[3] in order of performance that can be used to determine the disposition of a null hypothesis: (1) state the null hypothesis (H_o), (2) choose a statistical test for testing H_o [See the Decision Model], (3) specify a significance level (α) and a sample size (n), (4) find / assume the sampling distribution of the statistical test under H_o, (5) define the region of rejection, and

[3] The procedures were adapted from Siegel (1956). The definitions related to hypothesis testing were adapted from Siegel (1956), Mann (1992) and Elmore et al. (1997).

(6) compute the value of the statistical test, using the data obtained from the sample(s). If the computed value is in the region of rejection, reject H_o. If the computed value is outside the region of rejection, H_o cannot be rejected at the chosen level of significance.

Definitions: *Significance level* of a test is the probability of rejecting H_o when in fact it is true. The size of the rejection region depends on the value assigned to α. The commonly used values of α are 0.01, 0.05, and 0.10. In social science research, the most common value of α is 0.05.

Definition: The *critical value* or critical point is the place on a curve at which the rejection and nonrejection regions are divided.

Decision Model

The Decision Model is a chart designed to facilitate a decision about which statistical procedure is most appropriate in the analysis of a set of data. (See Figure 1) A few questions about a set of data can lead quickly to a decision about which statistical procedure is most appropriate. The questions are: (1) What type of measurement scale produced the data -- nominal, ordinal, or interval/ratio scale? (2) Were the data related or independent? Was there an absence of association between observations or were the observations related as in a pretest/posttest of the same group? (3) How many groups were measured? (4) How many measurements were involved? By answering these questions and following the Decision Model downward, the appropriate statistical procedure can be identified.

The Matrix view of the Decision Model provides another quick reference in determining the appropriate procedure. (See Table 2) The variables are in the left column; the procedures of choice according to the measurement scale are in the other three columns.

Table 2: Matrix View of the Decision Model

Groups/Measures	Nominal	Ordinal	Interval/Ratio
1 Group/ 1 Measure		Descriptive / Detailed One Variable	Descriptive / Detailed One Variable
Independent Data 2 Groups/ 1 Measure >2 Groups/1 Measure		 Mann-Whitney U Kruskal-Wallis	 Independent Group t-Test One Way ANOVA
Related Data 1 Group/2 Measures (Relation) 1 Group/2 Measures (Gain/Loss) 1 Group/>2 Measures		 Spearman Rank Wilcoxon Signed Rank Friedman	 Pearson Product Moment (Fisher t-Test) Paired t-Test
1 Group(% Data)/ 1 Group (Raw Data) 2 or >2 Groups Test of Independence	X^2 Goodness of Fit X^2 Test of Independence		
	Conditions: 1. Data is in categories and category frequencies are analyzed.	Conditions: 1. Data is ranked and ranks are analyzed.	Conditions: 1. Assume random sample 2. Assume normal curve

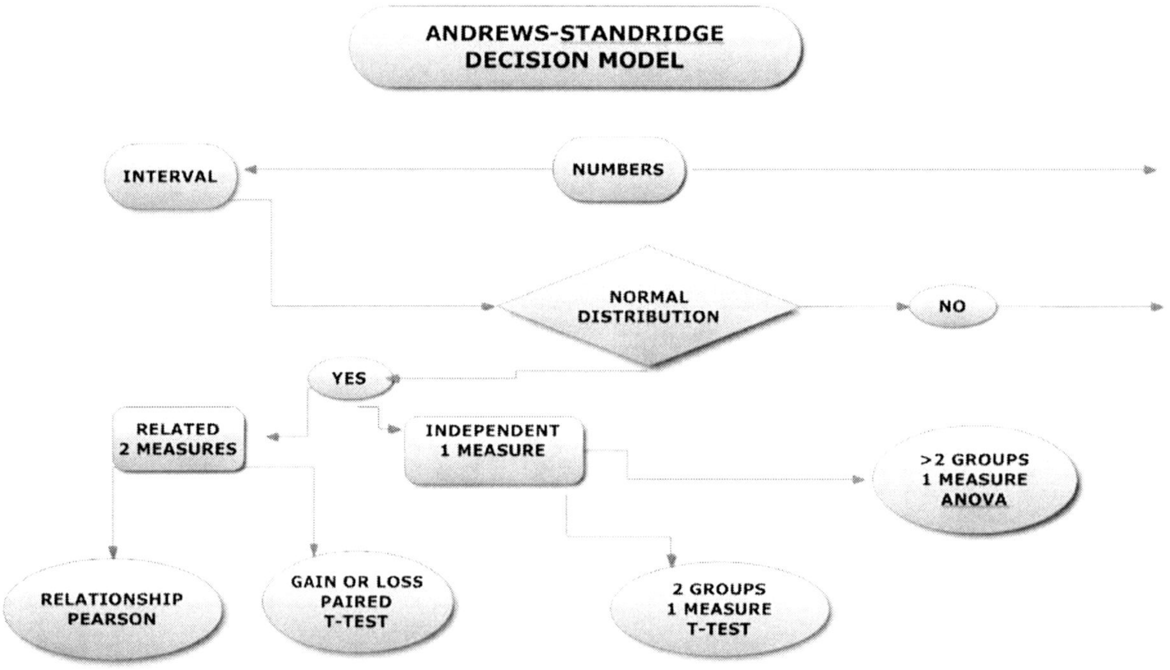
ANDREWS-STANDRIDGE
DECISION MODEL
INTERVAL
NUMBERS
NORMAL
DISTRIBUTION
NO
YES
RELATED
2 MEASURES
INDEPENDENT
1 MEASURE
>2 GROUPS
1 MEASURE
ANOVA
RELATIONSHIP
PEARSON
GAIN OR LOSS
PAIRED
T-TEST
2 GROUPS
1 MEASURE
T-TEST

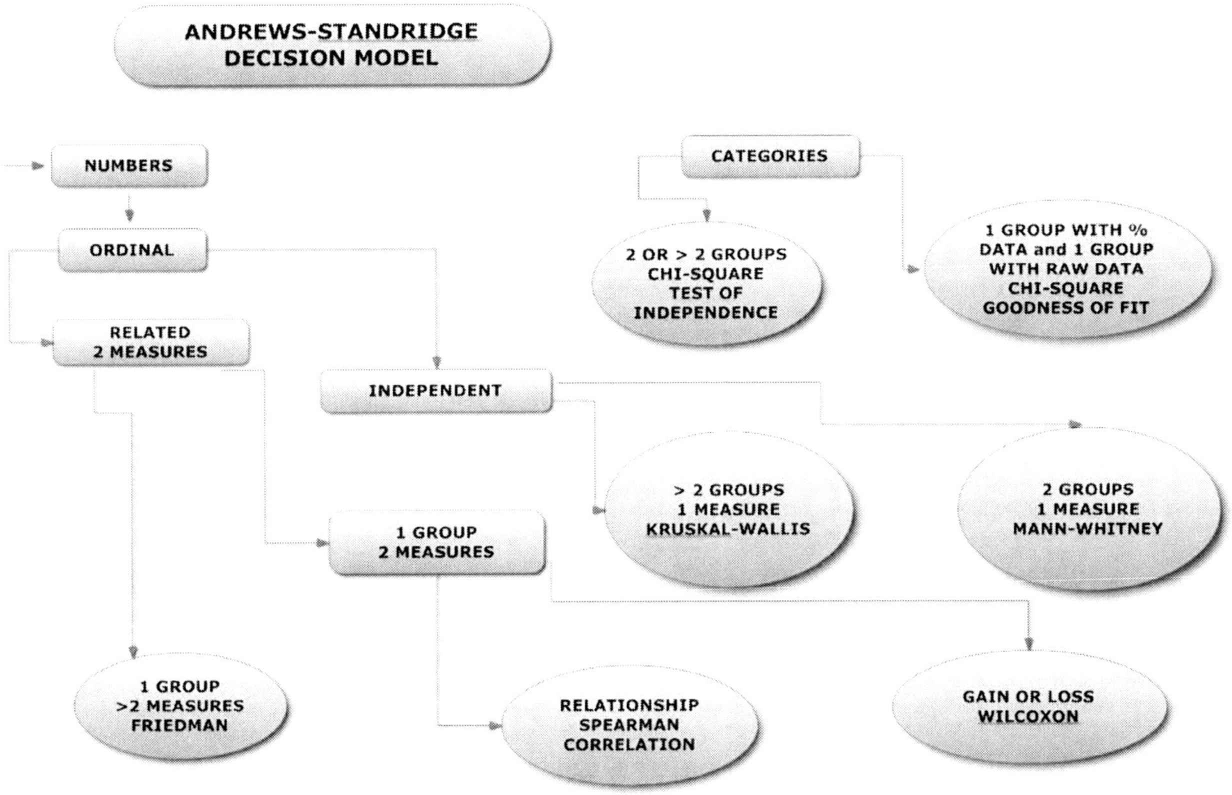
ANDREWS-STANDRIDGE
DECISION MODEL
NUMBERS
ORDINAL
RELATED
2 MEASURES
INDEPENDENT
1 GROUP
2 MEASURES
CATEGORIES
2 OR > 2 GROUPS
CHI-SQUARE
TEST OF
INDEPENDENCE
1 GROUP WITH %
DATA and 1 GROUP
WITH RAW DATA
CHI-SQUARE
GOODNESS OF FIT
> 2 GROUPS
1 MEASURE
KRUSKAL-WALLIS
2 GROUPS
1 MEASURE
MANN-WHITNEY
1 GROUP
>2 MEASURES
FRIEDMAN
RELATIONSHIP
SPEARMAN
CORRELATION
GAIN OR LOSS
WILCOXON

Exercise

1. Classify each of the following as nominal, ordinal, interval, or ratio level data.

    1) Horsepower of motorcycle engine
    2) Rating of newscasts in Houston as poor, fair, good, excellent
    3) Temperature of outdoors in degrees of Celsius
    4) Time required by drivers to complete a course
    5) Salaries of cashiers of Day-Night convenience store
    6) Political party preference of faculty members
    7) Age of students enrolled in a martial arts course
    8) Weight of beef cattle fed a special diet
    9) Ranking of contestant in a beauty contest
    10) Number of pages in the telephone book for Nashville

2. In the statements that follow, which are inferential statistics and which are descriptive.

    1) The average price of homes sold in Rhea County during the week of June 6-13 was $83,564.
    2) According to the Census Bureau, 20% of all Americans workers carpool to work.
    3) The National Eye Institute halted a clinical trial on eye surgery, calling it ineffective and possibly harmful to vision.
    4) The Gallup Poll says 1 of every 20 Americans is a member of a Country Club.
    5) Drinking decaffeinated coffee can raise cholesterol levels by 7%.
    6) According to the Court Administrator in Nashville, 10% of the civil actions in 1998 were decided in less than 6 months.
    7) Cigarettes were associated with 29% of the 4,470 civilian fire deaths in 1989.

3. Critique the statement, "statistics is a language."

4. Compare, contrast and illustrate the four functions of statistics – summarizing, describing, generalizing, and experimenting.

Using This Book

If the following information is committed to memory at the mastery-level of learning (instant, total recall), a student using this book will be able to conduct statistical research using the Decision Model to determine which statistic to use. Each chapter then can be used to produce a p-value for any hypothesis.

The Four Questions

1. Type of data?
2. Related or Independent data?
3. How many measures?
4. How many groups?

The Four Kinds of Data (NOIR)

Kind	**Definition**	**Examples**
Nominal	Naming or Categories	Gender or Political Party
Ordinal	Order or Rank	Horse Race or Likert-type scale
Interval	The intervals are Equal	Height, Weight, Distance, Time
Ratio	Interval data with a zero	

Example

H_o There was a difference in GPA between male and female students.

1. Type of data? Interval
2. Related or Independent? Independent (only one measure)
3. Number of measures? One (GPA)
4. Number of groups? Two (males and females)

Using the Decision Model, the correct procedure is the t-Test (Chapter Six).

CHAPTER 2

CENTRAL TENDENCIES

Central tendency is any of several measures of general magnitude of the numbers in a distribution. Measures of central tendency describe a distribution in words without needing tables or graphs. The most common measures of central tendency are the mean, the median, and the mode.

Definition: A *frequency distribution* is a listing of the frequency of observations in each category, usually as a table and in order of magnitude.

Mean

A mean is a number indicating the central value of a group of observations. The mean is sometimes known as the *average.* When the mean of a distribution is subtracted from each score in that distribution and the differences are added, the sum will be zero.

A mean serves two important purposes. First, a mean is a shorthand description of a mass of quantitative data obtained from a sample. It is more meaningful and economical to let one number stand for a group than to try to note and remember all numbers. Secondly, a mean describes indirectly but with some accuracy the population from which a sample is drawn (Guilford 1965).

Ungrouped Data

To get the mean of ungrouped data, sum all of the measurements and divide by the number of measurement. For a sample, the following is the procedure for a mean of ungrouped data.

$$\bar{x} = \frac{\sum x}{n} \qquad (2.1)$$

x = each of the measurements

Σ = the sum of

$\bar{x}$ = arithmetic mean

n = the number of measurements

For a population, the formula for a mean of ungrouped data is.

$$\mu = \frac{\sum x}{N} \qquad (2.2)$$

μ = the arithmetic mean of population

Σ = the sum of

x = each of the measurements

N = number of measurements

Note that $\bar{x}$ is the symbol for the mean of a sample and μ is the mean for a population. In the same way, *n* is the symbol for the number in a sample and *N* is the symbol for the number in a population.

A class of 10 students scored 69, 71, 77, 78, 79, 81, 84, 86, 91, and 94 on a test [See Table 3]. Note that the numbers are distributed in ascending order from the lowest score to the highest score. The numbers could have been distributed in descending order with 94 at the top.

Table 3 -- Ungrouped Data Mean

69	78	84	94	Σx = 810
71	79	86		n = 10
77	81	91		$\bar{x}$ = 81

Grouped Data

When the data are grouped for whatever purpose, it is more convenient to apply a special formula to determine the mean.

$$M = \frac{\Sigma f X_C}{N} \qquad (2.3)$$

M = mean of grouped data

$\sum$ = the sum of

f = number of cases within an interval

fX = the frequency times the X value for the interval

X_c = midpoint of a class interval

Table 4[4] -- Grouped Data Mean

Interval	X_C Midpoint	f	fX_C
75-77	76	1	76
72-74	73	2	146
69-71	70	3	210
66-68	67	4	268
63-65	64	4	256
60-62	61	6	366
57-59	58	5	290
54-56	55	4	220
51-53	52	3	156
48-50	49	2	98
45-47	46	1	46
TOTAL		N = 35	ΣfX_C = 2165 M = 61.86

$$M = \frac{\Sigma f X_C}{N} = \frac{2165}{35} = 61.86 \qquad (2.3)$$

[4] The Frequency Table for Grouped Measurements was developed by Maurice H. Field (1996, 18)

Median

Ungroup Data

The median for ungrouped data is the middle value of a set of numbers arranged in order of magnitude. The median is the point on a scale of measurement above which are exactly half the cases and below which are the other half. Note that a median is a point and not a score or any particular measurement (Guilford 1965). The median is often the best measure of central tendency if the scores are skewed. For small samples, the median is usually the middle value of an odd number of scores and half way between the two middle numbers of an even number of scores.

Odd Number of Scores with No Duplication in the Region of the Median -- To calculate the median for ungrouped data with no duplication of scores in the region of the median, rank-order the scores from high to low. Use the following formula to find the value corresponding to rank order position of the median in the array of numbers.

$$Mdn(Ungrouped) = \frac{n+1}{2} \tag{2.4}$$

Among the ten numbers in the mean example, the median is half way between 79 and 81 (See Table 3). The median is 80, half way between 79 and 81. Half of the numbers are above 80 and half are below 80.

$$Mdn = \frac{10+1}{2} = 5.5 \tag{2.4}$$

Odd Number of Scores with Duplication in the Region of the Median

--To calculate the median for ungrouped data with an odd number of values and a duplication of values in the region of the median, rank-order the scores from high to low. Find the value corresponding to rank order position with the same formula as used for ungrouped data with no duplication of scores in the region of the median. If the value is duplicated, substitute the following formula for the median.

$$Mdn = l + \frac{\frac{N}{2} - n_b}{n_w} \qquad (2.5)$$

where

l = the lower exact limit of the score interval containing the median

N = the total number of scores

n_b = the number of scores falling below the lower real limit

n_w = the number of scores falling within the interval containing the median

Nine values are presented for calculation of the median as follows: 7, 7, 6, 6, 5, 5, 5, 4, 3. There are an odd number of values and duplication in the region of the median. The calculation of the median is as follows.

$$Mdn = 4.5 + \frac{\frac{9}{2} - 2}{3} = 4.5 + \frac{4.5 - 2}{3} = 4.5 + .83 = 5.33 \qquad (2.5)$$

Even Number of Values with No Duplication in the Region of the Median – To calculate the median with no duplication of values in the region of the median, (1) rank-order the values from high to low. (2) Find the value corresponding to rank-order position N ÷ 2. (3) Locate the value corresponding to rank-order position $\frac{(N+2)}{2}$. (4) If the above two values are different, interpolate to find the mid-point between them, which will be the median.

$$Mdn = \frac{(L+H)}{2} \qquad (2.6)$$

where

L = the lower value

H = the higher value

Six values, and even number of values, are presented for calculation of the median as follows: 13, 12, 8, 6, 5, 3. Rank order position N ÷ 2 = 6 ÷ 2 = 3. The value corresponding to rank-order position $\frac{(6+2)}{2} = 4$. The two scores are different; interpolate to find the mid-point. $Mdn = \frac{(6+8)}{2} = 7$

Even Number of Values with Duplication in the Region of the Median – To calculate the median with an even number of values with duplication in the region of the median, (1) rank-order the scores from high to low. (2) Find the score corresponding to rank-order position N ÷ 2. (3) Locate the value corresponding to rank-order position $\frac{N}{2}$. (4) If the above two scores are the same, use the formula

$$Mdn = l + \frac{\frac{N}{2} - n_b}{n_w} \qquad (2.7)$$

where

l = the exact lower limit of the score interval containing the median

N = the total number of scores

n_b = the number of scores falling below the lower real limit

n_w = the number of scores falling within the interval containing the median

Ten values, and even number of values, are presented for calculation of the median as follows: 40, 38, 36, 36, 36, 36, 36, 24, 23, and 10. L = 36.

$$Mdn = 36 + \frac{\frac{10}{2} - 3}{5} = (36 + \frac{2}{5}) = (36 + .40) = 36.4$$

Grouped Data

As with ungrouped data, the median of grouped data is that point on the measuring scale in which half of the observations are above and half of the observations are below. In Table 4 above, there are 35 cases, so the median is that point on the scale above which there are 17 cases and below which there are 17 cases. Counting frequencies from the bottom upward, 1 + 2 + 3 + 4 + 5 = 15. The next number is 6, which would make 21, but only 3 are needed. The median lies somewhere in the interval of 60-62. The exact interval limits are 59.5 to 62.5. This means that 6 cases in this interval are spread over the distance from 59.5 to 62.5. By interpolating within this range, it can be determined how far from 59.5 is needed in order to include the three cases needed for the 18th case, which is the median. To get the 3 cases, 3 of the six cases are needed. The total units are three, so 3/6 or 1/2 of 3 is 1.5. Adding 1.5 to 59.5, the median is 61.

The procedure for interpolation of a median from the bottom up is:

$$Mdn = l + \left(\frac{\frac{N}{2} - F_b}{f_p} \right) i \qquad (2.8)$$

l = exact lower limit of the class interval containing the median

F_b = sum of all frequencies below l

f_p = frequency of the interval containing the median

i = the size of the class interval

Mode

The mode is the single value, which occurs most frequently and is usually easier to identify with grouped data than with ungrouped data. With ungrouped data, the mode is that measurement that occurs most frequently. Usually the mode is somewhere near the center of the distribution, and in a normal distribution coincides with the mean and the median. In a distribution of grouped data, the crude mode is the midpoint of that class interval having the greatest frequency (Guilford 1965).

In most instances a distribution is unimodal, that is, has only one mode. It is possible for a distribution to have more than a single mode. If two nonadjacent values in a group of numbers have the same frequency and their frequency is greater than for any other values in the group, then two modes exist and the distribution is considered bimodal. Multimodal distributions have many modes.

Determining Skewness

Not all distributions are normal. Sometimes a distribution is skewed negatively; sometimes positively. When the mean is larger than the median, the distribution is positively skewed and the curve will lean to the left. When the mean is smaller than the median, the distribution is negatively skewed and the curve will lean to the right. The mean is affected by the size of the numbers and is pulled in the direction of the extreme values. The median is not influenced by the size of extreme values.

Definition: Skewness is the degree of asymmetry of a distribution.

Central Tendency Uses

Compute the mean when (1) the greatest sampling stability is wanted, (2) other computations, such as finding measures of variability follow, (3) the distribution is symmetrical about the center, and (4) the center of gravity of a sample is desired.

Compute the median when (1) there is not sufficient time to compute mean, (2) distributions are markedly skewed, (3) there is interest in knowing whether cases fall within the upper or lower halves of the distribution and not particularly in how far they are from the center point, and (4) an incomplete distribution is given.

Compute the mode when (1) the quickest estimate of central value is wanted, (2) a very rough estimate of central value will do, and there is interest in knowing the most typical case. (Guilford, 1965)

Exercise

1. The following distribution was the number of errors on a statistics exam. One participant missed 10, 2 missed 9, 1 missed 8, 1 missed 7, 3 missed 6, 1 missed 5, and 1 missed 4.

    1) Using Microsoft Word or Excel, develop a table with four columns that is no more than 3 inches wide, and is single-spaced. Center the headings at the top of each column, two rows deep and entitle Column 1 *Errors* and place the symbol X on the second line. Column two should be entitled *Frequency* and the symbol *f* should be on the second line. Column 3 should be labeled *Midpoint* on the first line and X_c on the second line. Column 4 should be labeled on the second line only with $f(X_c)$.

2) Calculate the mean, median, and mode manually using the procedures described in chapter 2. Make the bottom row a single cell and center the calculations with correct symbols in the bottom row of the table.
3) Enter the distribution of values and use Minitab® and Calc (on the toolbar) to calculate the mean, median, and mode. Is the distribution of the exam errors symmetric, positively skewed, or negatively skewed?

2. If all scores in the distribution are multiplied by 2 and then 3 is subtracted from each, what is the mean of the new distribution of scores?

CHAPTER 3

DISTRIBUTION

Normal Curve

A normal distribution is a symmetric, unimodal, bell-shaped probability distribution, used as the theoretical model for the distribution of a large number of physical and psychological variables (Yaremko 1982). The normal distribution is also known as the Gaussian Distribution[5]. The z score is the unit of measurement on the horizontal axis of a normal curve.

A normal curve has at least 5 basic characteristics. (1) The normal distribution curve is bell-shaped. (2) The curve is symmetric about the mean. One-half of the total area under a normal curve lies on the left side and one-half lies on the right side of the mean. When graphed the values tend to form a continuous curve in the form of a bell.[6] (3) The two tails of the curve extend indefinitely in both directions without touching or crossing the horizontal axis. Beyond 3 standard deviations (3σ) the curve may come so close to the axis that the curve beyond these points in both directions can be viewed as virtually zero. (4) The mean, median, and mode are the same. (5) The total area under the curve is 1.0.

---

[5] Huguenot refugee Abraham de Moivre discovered the normal distribution in 1733 as an approximation to the binomial distribution when the number of trials is large. The distribution is more usually associated with the name of Karl Friedrich Gauss (1777-1855) who derived it in 1809 in his *Theoria motus corporum coelestium*, as the law of errors of observations. (Bulmer 1979)

[6] Sir Frances Galton (1822-1911), English scientist, took systematic measurements of numerous psychological and physical variables on very large numbers of individuals. He found that the distributions of many of the measurements were an approximate normal distribution.

The normal curve determines the probability of an event when the event class is known or assumed to be normally distributed. There are two kinds of frequency distributions (1) empirical frequency distributions, which have real values or scores, and (2) theoretical frequency distributions, which are generated by some formula. Included in theoretical frequency distributions are the theoretical normal distribution and the normal curve.

Central Limit Theorem

The importance of the normal distribution lies in the Central Limit Theorem.[7] The Central Limit Theorem states that the sum of a large number of independent random variables will usually approximate a normal distribution regardless of the individual distributions. The central limit theorem does not apply unless the factors are independent and additive; the requirement of additivity is particularly important. Another important application of the Central Limit Theorem is to the binomial and Poisson distributions.

Another application of the theorem lies in the theory of sampling distributions, which usually become approximately normal in large samples even though the underlying distribution may be far from normal. Sampling distribution tends to the normal form as *n* increases regardless of the form of the underlying distribution. The assumption of normality breaks down only if *n* is very small or if the underlying distribution is highly abnormal.

---

[7] Laplace first proved the theorem in 1812 but the significance of the theorem was not realized until Quetelet demonstrated in 1835 that the normal curve described the distribution not only of errors of measurement but also of biological variables such as human height. It was soon discovered that many other biometrical variables followed this distribution; it was at this time that it acquired the name normal distribution. (Bulmer 1979)

Mean and Standard Deviation

The mean and standard deviation of the distribution determine the exact shape of the curve. The mean locates the peak or center of the curve and the standard deviation determines the spread or width of the curve. A small standard deviation will be more peaked; a large standard deviation will be flatter.

Statistical Selection

The shape of the distribution is important in the selection of a statistical test. Most parametric procedures require a normal or near normal distribution of data. If the data does not conform to the requirements for normal distribution, nonparametric statistical procedures are used.

Base Line

The base line is the area under the curve containing 100% of the cases in a distribution. Between the mean and one standard deviation[8] from the mean, 34.13% of the cases or frequencies are included. Between the mean and 2 standard deviations, 47.72% of the cases are included. Between the mean and 3 standard deviations, 49.87% of the cases are included. This is sometimes referred to as the Empirical Rule. By including the percentage on each side of the mean (1) the number of cases between $-1\ \sigma$ and $+1\ \sigma$ = 68.26%, (2) the number of cases between $-2\ \sigma$ and $+2\ \sigma$ = 95.44%, and (3) the number of cases between -3σ and $+3\sigma$ = 99.74%.

[8] σ is the symbol for the standard deviation of a population; s is the symbol for the standard deviation of a sample.

An important function of the normal curve is clarification and interpretation of standard deviation. On a normal curve the standard deviation measures and graphically displays variability or range of frequencies. Two sets of frequencies may have the same mean and a significantly different standard deviation because the range of scores varied widely. A normal distribution with a mean of 25 and a standard deviation of 5 is symbolized by *n*(25, 5).

Probability

Probability is a quantitative statement about the likelihood of the occurrence of an event. In other words, the probability associated with an event is the number of times that event may occur relative to the total number of times any event can occur. The terms *likelihood* or *chance* are synonyms for probability. A probability of zero refers to an event that will never occur; a probability of one refers to an event that will always occur; values in between refer to the likelihood of are event out of several that could occur. Probability is used to measure uncertainty.

$$p_{(e)} = \frac{n_{(e)}}{n_{(t)}} \qquad (3.1)$$

p(e) = the probability of a desired event

n(e) = the number of times an event can or does occur

n(t) = is the total number of possible experiments

Because n(e) cannot possibly be larger than n(t), p(e) will never be greater than one nor less than zero.

$$p_{(e)} = \lfloor p_{(e1)} \rfloor \times \lfloor p_{(e2)} \rfloor \qquad (3.2)$$

$p(e_1)$ = the probability of one simple event

$p(e_2)$ = the probability of another simple event

If the compound event is made up of more than two independent simple events, then all probabilities of simple events are multiplied together. Probability decreases when traveling the base line of a normal curve in either direction away from the mean. Of the total frequencies 68.26% fall between -1σ and $+1\sigma$ from the mean. In other words the probability of any given raw score falling within this interval is 68 chances in 100. Probability associated with the areas under a normal curve is always given relative to 100% which is the entire area under the curve. For this reason and as a standard, probability is always the number of times any given event can occur out of 100. Probability is usually expressed in decimals as a proportion. For example, P = .50 or 50% of getting heads in one flip of a coin. Expressed as a proportion, probability always ranges between 0 to 1 (Levin 1983).

Addition Rule

The probability of obtaining any one of several outcomes equals the sum of the separate probabilities. For example, the probability of drawing three specified cards from a deck of 52 cards would be 1 chance in 52 for each card or 3 chances in 52 of drawing one of the three cards. The probability expressed a proportion of 100 is .06 [P = .06] (Levin 1983).

Multiplication Rule

The probability of obtaining a combination of mutually exclusive outcomes is equal to the product of the separate probabilities (Levin 1983). Using coin flips as an example, show the probability of flipping the coin twice and getting two heads or flipping three times and getting three tails.

Two Flips: $p(e) = [p(e_1)]\,[p(e_2)] = (.5)\,(.5) = .25$ (3.2)

Three Flips: $p(e) = [p(e_1)]\,[p(e_2)]\,[p(e_3)] = (.5)\,(.5)\,(.5) = .125$

Many important decisions are made on the basis of the probability that an event will or will not occur. Probability is important in many statistical analyses. One word of caution -- a thousand probabilities do not make one fact.

Standard Scores

A standard score describes a number in a distribution relative to other numbers in the distribution by measuring the deviation of the number from the mean of the distribution in standard deviation units. It is not statistically permissible to compare or average raw scores from different tests unless the distributions have the same mean and standard deviation. In converting raw scores to standard scores, the raw scores are changed to distributions which have a mean of zero and a standard deviation of one. Changing a distribution of raw scores does not change the shape of the distribution of scores. Changing a distribution of raw scores to standard scores is called data transformation. Standard scores can be added, averaged, and combined with equal weight; if the raw score distribution has the same form.

The most common form of a standard score is a *z-score.* A *z-score* is defined as the difference between the raw score and the mean divided by the standard deviation. A z-score has four properties (1) the mean of the z-scores is zero, (2) the variance of the z-scores is one, (3) the sum of the z-scores is zero, and (4) the sum of the squared z-scores is N. (Elmore et al. 1997)

The procedure for calculating a z-score is as follows:

$$z = \frac{x - \bar{x}}{\sigma} \tag{3.3}$$

x = raw score

$\bar{x}$ = the raw score mean

σ = standard deviation of the raw score

Any set of measurements with a known $\bar{x}$ and a standard deviation can be transformed into a z-score to show the deviation of the raw measurement from the mean, measured in standard deviation units. Positive standard scores are above the mean of the group. Negative standard scores are below the mean of the group. In the exercise below, a standard score of 1.53 is 1.53 standard deviations above the mean of the group and a standard score of
–2.14 is 2. in standard deviations below the mean of the group. (See Table 5)

Since the standard score is partially determined using the mean and the standard deviation, an average of standard scores can easily be determined by adding the various standard scores. For example, the five z-scores, -1.5, -1, +1, +1.5, and +2, can be averaged by adding: +2.

Table 5 -- Data Transformed to Standard Scores

Raw Scores	*f*	$x - \bar{x}$	(z) $\frac{x - \bar{x}}{s}$
39	1	11.76	2.06
36	1	8.76	1.53
35	2	7.76	1.36
34	1	6.76	1.18
30	3	2.76	48
29	5	1.76	.31
27	6	-.24	-.04
26	4	-1.24	-.22
25	2	-2.24	-.39
23	1	-4.24	-.74
20	2	-7.24	-1.27
17	1	-10.27	-1.79
15	1	-12.24	-2.14
12	1	-15.24	-2.69
	N = 33	$\bar{x} = 27.24$ s = 5.71	

Practical Use of the Normal Curve

The normal curve is used to determine percentage, percentile rank, proportion, or a probability within a population. Three steps are involved in the use of the Normal Curve (1) convert the raw score(s) to z-score(s), (2) from the Normal Curve Table determine the area of the curve between the mean (μ) and the z-score, and (3) perform whatever arithmetic operation is needed to answer the question.

t-Distribution

The t-distribution is used when the sample size is less than 30 and the variable is normally or approximately normally distributed to determine the critical value when comparing means and proportions. The t-distribution is similar to the standard normal distribution in the following ways: (1) it is bell-shaped, (2) it is symmetrical about the mean, (3) the mean, median, and mode are equal to 0 and are located at the center of the distribution, (4) the curve never touches the x axis. The t-distribution differs from the standard normal distribution in the following ways: (1) the variance is greater than 1, (2) the t-distribution is a family of curves based on the degrees of freedom or sample size, and (3) as the sample size increases, the t-distribution approaches the standard normal distribution. Gosset derived the t-distribution by selecting small random samples of measurements taken from a population of incarcerated criminals. He measured the lengths of one finger.

F-Distribution

The F-distribution is used to determine the critical value when comparing parametric variances. The F-distribution is a family of curves that compare means based on the degrees of freedom of the variance of the numerator and the degrees of freedom of the variance of the denominator. The value of F is always positive or zero and is positively skewed.

Chi-Square Distribution

The chi-square distribution is used to determine the critical value when comparing nonparametric variances and standard deviations. In statistics, the variance and standard deviation of a variable are as important as the mean. The chi-square distribution is a family of curves based on a single number of degrees of freedom. A chi-square variable cannot be negative, and the distributions are positively skewed. At about 100 degrees of freedom, the chi-square distribution becomes somewhat symmetrical. The area under each chi-square distribution is equal to 1.00.

Hershel formulated the chi-square distribution with 2 degrees of freedom in 1869 while studying the accuracy of shooting arrows at a target. Other mathematicians have since contributed to the development.

Exercise

1. Niblick McDivot, star golfer of Pancreas University was informed that he must score at least as high as the average student in his classes in order to tee off in the big intercollegiate tournament. His grades were given standard scores as follows: Science 1.48, Mathematics .96, English -.54, History –1.20 and Art .82. Does Niblick get to play in the big tournament?
2. In a normal distribution of scores, what percent of the total area falls (1) between –1 standard deviation and +1 standard deviation, (2) between –2 standard deviations and +2 standard deviations, and (3) between –3 standard deviations and +3 standard deviations?
3. Given a normal distribution of raw scores in which the mean is 7.5 and the standard deviation is 1.3, express each of the following raw scores as a z score: (1) X = 8.0, (2) x = 6.0, (3) X = 5.3, (4) X = 10.2, and (5) X = 7.5.

CHAPTER 4
MEASURES OF VARIABILITY

Measures of variability are indexes of how values are scattered around the center of a distribution. Variability is also known as spread, width, or dispersion. The most common values to indicate variability are range[9], the semi-interquartile range[10], the average deviation[11], variance[12] and the standard deviation[13].

General Example

The following example points out the need for measures of variability. A man named Quick Sand hired ten workmen to shovel sand while he was away on business. At the end of the day, the ten workmen had shoveled 500 buckets of sand. The mean number of buckets shoveled was 50; the median was 50 and the mode was 50. While these central values give some information, there are unanswered questions. For instance, were there any loafers among the shovelers? Were there any stars? Quick Sand needed more information than was provided by central tendencies to adequately evaluate the performance of the ten shovelers.

The range of the number of buckets of sand shoveled by the ten workmen would give some additional information. In other words, what was the true upper limit and what was the true lower limit? In addition to identifying the high and the

---

[9] The symbol for range is *R*.
[10] The symbol for interquartile range is *Q*.
[11] The symbol for average deviation is *AD*.
[12] The symbol for the variance of a population is σ^2 and s^2 for a sample.
[13] The symbol for standard deviation of a population is σ; the symbol for a sample is s.

low production, the range can also be used to establish intervals for grouped data.

Other information about the shovelers was gained by examining the distance of the total production of each workman from the points of central value. The distance of the values from the points of central value is known as deviations. The average of the deviation is one index; standard deviation is an even more commonly used index.

The ten men employed to fill buckets with sand shoveled 45, 46, 47, 48, 50, 50, 52, 53, 54, and 55 buckets of sand. The deviations from the mean of 50 were –5, -4, -3, -2, 0, 0, 2, 3, 4, and 5. The sum of the deviations was 28; the average deviation was 2.8. The standard deviation, based on the squared deviations from the mean was 3.4641. The semi-interquartile range or one-half of the range between the 25th and 75th percentile was exactly 3.25 on both sides of the mean. Although the sample was small, the distribution of the data are approximately normal.

Range

A quick way to measure variability is to find the range or *R*, which is the difference between the highest and lowest values in a distribution of numbers. The big disadvantage of range is the dependency upon just two extreme values, the highest and the lowest numbers. All other values in a distribution are ignored. By changing either the highest or lowest number, the range can fluctuate dramatically. Any measure that is so affected by the value of only one change cannot give a precise measurement of variability. At best, range must be considered as only a preliminary or very rough index (Levin 1983).

Semi-Interquartile Range

The semi-interquartile range tells the amount and direction of skewness in a distribution of numbers. Skewness is the extent to which a distribution deviates from symmetry. In a positively skewed bell shaped curve, the tail of the frequency distribution points to the right and in a negatively skewed curve, the tail points to the left (Yaremko 1982).

The first quartile or Q_1 is a point one-fourth the way up a distribution of numbers, which are ranked in order of magnitude. The third quartile or Q_3 is found in a similar manner except the point begins one-fourth way down from the top. The quartiles are the division point between the quarters.

The interquartile range is the distance from Q_1 to Q_3. The semi-interquartile range is one-half of the interquartile range.

$$Q = \frac{Q_3 - Q_1}{2} \qquad (4.1)$$

Q_1 = first quartile

Q_3 = third quartile

If the distribution is exactly symmetrical, the third and first quartiles will be the same distance from the median, and that distance is Q. When there is any skewness in the distribution, the two distances will be unequal (Guilford 1965).

The use of interquartile and semi-interquartile ranges are better than the use of range for descriptive purposes because they are functions of two values in a more central and stable part of the distribution than the two extreme values of the range. When the median is used as a measure of central tendency, the semi-interquartile range is the measure of variability used. Researchers often report two numbers, the Median minus *Q* and the Median plus *Q*.

Percentiles

The percentile rank of a value is a single number, which indicates the percent of cases in a distribution, which fall below a given value. Percentile rank values range from 1 to 99 and reference only the portion of the distribution, which is lower than a number. The median, which is the middle value of a set of numbers arranged in order of magnitude, is always the 50th percentile.

The median or other percentiles are sometimes difficult to find in a distribution of grouped data. The following procedure has been developed for computing the median or any of the desired percentiles for grouped data[14].

$$PRofX = \frac{100}{N}\left[cf + \frac{(X - L)f}{W}\right] \qquad (4.2)$$

PR = Percentile Rank

X = a raw score value from the distribution

N = total number of observations in the data set

cf = cumulative frequency of interval below the interval containing the raw score value

L = lower real limit of the interval containing the raw score value

f = frequency of scores or values in the interval containing the score point

W = width or size of interval

[14] See *Basic Statistics*, page 34 for a procedure for determining a percentile with ungrouped data. The following procedure and calculation steps come from *Basic Statistics* (Elmore et al. 1997).

Calculation of a Percentile Rank

To calculate a percentile rank (1) locate the interval containing the score point, (2) determine the lower real limit of the interval, (3) find the frequency of scores in the interval and the width of the interval. (4) find the cumulative frequency of the interval below the interval containing the score point, and (5) calculate the percentile rank.

Average Deviation

All values in a distribution of numbers deviate from the mean except the numbers that coincide with the mean. The procedure for finding a deviation, which has the symbol *x* is as follows:

$$x = X - M \qquad (4.3)$$

x = a deviation of a measurement from the mean

X = the original value or measurement

M = arithmetic mean

To find the average deviation, disregard the algebraic signs of the deviations and take into account only the size of the deviations. This is known as averaging absolute values. In the average deviation, each measurement is utilized, but no greater emphasis is given to a wide deviation than to a small deviation from the mean (Field 1996). The average deviation provides a sound basis for understanding the nature of the standard deviation.

Variance

A measure of variability, which takes into consideration each measurement in the distribution and emphasizes the magnitude of the deviation from the mean, is the variance. The variance can be defined as the mean of the squared deviations from the distribution mean. The deviation score procedure[15] for calculating the variance of a population is as follows:

$$\sigma^2 = \frac{\Sigma\left(x - \bar{x}\right)^2}{N} \tag{4.4}$$

σ^2 = population variance

Σ = Sum of

$\left(x - \bar{x}\right)^2$ = square of the deviations from the mean of the population

$\bar{x}$ = mean of the population

N = size of the population

The sum of the squared deviations reflects not only the variability of the group, but also the size of the group, the variance of a small group, or a sample taken from a population. The procedure for calculating the variance of a sample is as follows:

$$s^2 = \frac{\Sigma\left(x - \bar{x}\right)^2}{n - 1} \tag{4.5}$$

s^2 = variance of a sample

Σ = Sum of

$\left(x - \bar{x}\right)^2$ = square of the deviations from the mean of the sample

$\bar{x}$ = mean of the sample

n = size of the sample

---

[15] A second formula for calculating the variance and standard deviation is called the raw score formula.

Standard Deviation

Standard deviation is a measure of the variability or dispersion in a set of numbers that provides an indication of the average amount by which the scores deviate from the mean of the distribution. Standard deviation is the square root of the variance. The deviation score procedure for calculating the standard deviation of a population is as follows:

$$\sigma = \sqrt{\frac{\Sigma(x-\mu)^2}{N}} \tag{4.6}$$

σ = standard deviation

x = deviation from the mean of the population

μ = mean of the population

N = size of the population

The deviation score procedure for calculating the standard deviation for a sample is as follows:

$$s = \sqrt{\frac{\Sigma(x-\bar{x})^2}{n-1}} \tag{4.7}$$

s = standard deviation

x = deviation from the mean of the sample

$\bar{x}$ = mean of the sample

n = size of the sample

Computing Standard Deviation

To calculate standard deviation (1) rank the numbers in order of magnitude from the largest to the smallest; (2) find each deviation from the mean; (3) square each deviation, finding $(x-\bar{x})^2$; (4) *sum* the squared deviations, finding $\Sigma(x-\bar{x})^2$; (5) divide the squared deviations by *N,* finding $\Sigma(x-\bar{x})^2/N$, which is the variance; (6) extract the square root of the results. This is the standard deviation.

General Example

Following is an example of the procedure for calculating the standard deviation for a small, ungrouped sample. The ages of a population of 15 temporary workmen are 50, 48, 47, 46, 45, 45, 20, 20, 19, 19, 19, 19, 19, 19, and 18. The employer wants to know the mean and standard deviation of the ages. The ages are ranked in order of magnitude from largest to the smallest. The mean age is 30.2; the symbol for the mean is $\bar{x}$. A table is a convenient tool with which to calculate, square, and sum the deviations of each age from the mean. The symbol for a deviation is $(x-\bar{x})$. Each deviation is squared finding $(x-\bar{x})^2$. The sum of the squared deviations, $\sum(x-\bar{x})^2$ = 2,788.4. The variance for a sample is the sum of the squared deviations divided by the number of ages minus 1, or $\frac{\sum(x-\bar{x})^2}{n-1}=\frac{2788.4}{14}=199.1714$. The standard deviation is the square root of the variance or $\sqrt{199.1714}$ = 14.11.

Had the ages been a population instead of a sample, the variance would be the squared deviations divided by the number of ages or

$\frac{\sum(x-\bar{x})^2}{N} = \frac{2,788.4}{15} = 185.89333$. The standard deviation is the square root of the variance or $\sqrt{185.89333} = 13.63$.

Table 6 – Standard Deviation Worksheet

Age (x)	Mean Age ($\bar{x}$)	x - $\bar{x}$	(x - $\bar{x}$)^2
18	30.2	-12.2	148.84
19	30.2	-11.2	125.44
19	30.2	-11.2	125.44
19	30.2	-11.2	125.44
19	30.2	-11.2	125.44
19	30.2	-11.2	125.44
19	30.2	-11.2	125.44
20	30.2	-10.2	104.04
20	30.2	-10.2	104.04
45	30.2	14.8	219.04
45	30.2	14.8	219.04
46	30.2	15.8	249.64
47	30.2	16.8	282.24
48	30.2	17.8	316.84
50	30.2	19.8	392.04
n = 14 R = 50-18 = 32	$\bar{x} = 30.2$		$\Sigma(x-\bar{x})^2$ = 2,788.4
$s = \sqrt{\frac{\sum(x-\bar{x})^2}{n-1}} = \sqrt{\frac{2788.4}{14}} = \sqrt{199.1714} = 14.11$			

MINITAB Example

The central tendencies and measures of variability for the ages in the General Example were analyzed by the MINITAB program using the following procedures: (1) The MINITAB program is opened. (2) The blue shaded, column heading box (C1) is clicked and AGE is entered. (3) Starting at row 1 the values are entered. (4) The STAT menu is clicked, then Basic Statistics and Display Descriptive Statistics. (5) In the Display Descriptive Statistics Window click C1

AGE and click Select; then click OK. The results appear in the viewer. Following is an example of the MINITAB output.

Table 7 -- MINITAB Descriptive Statistics / AGE Printout

Descriptive Statistics: AGE

Variable	N	Mean	Median	TrMean	StDev	SE Mean
AGE	15	30.20	20.00	29.62	14.11	3.64

Variable	Minimum	Maximum	Q1	Q3
AGE	18.00	50.00	19.00	46.00

Exercise

1. For the following sample and population distribution of numbers find the mean, median, mode, range, variance, and standard deviation: 26, 25, 24, 21, 20,18, 17, 14, 14, 12, 11, 10, 8, 4, 1. Either enter the distribution of numbers listed in #1 above into the MINITAB® Stats/Display Descriptive Stats/Statistics (check appropriate boxes) program and print the resulting analysis, or enter the column of numbers in Excel and determine the values using the functions in Excel.

CHAPTER 5

GRAPHICS

The two most direct ways to organize data is by creating tables and graphs. Developing tables usually precedes the creation of a graph. Turabian (1996)[16] provides guidelines for tables and guidelines for illustrations, which include charts, graphs, maps, drawings, photographs and paintings. There are several different types of tables and many kinds of graphs.

Tables

Frequency Table

Probably the simplest table to create is a frequency table. A frequency table lists the different categories of data and the associated frequency of occurrence. Frequency is a count of the number of subjects falling into the

Table 8 – Frequency Table

X	***f***
21	1
20	1
19	1
18	0
17	1
16	4
15	4
14	1
13	2
12	1
11	1
10	2
9	1

[16] Turabian (1996), Chapter 6, pages 87-103 are the guidelines for tables. Chapter 7, pages 104-115 are the guidelines for illustrations.

different categories. Frequency tables can be used to display nominal, ordinal, interval or ratio data.

Any category of frequencies can be divided by the total frequencies to obtain relative frequencies, which are in the form of a decimal value or a percentage. Relative frequency is independent of the size of the data set and is important when comparing two or more data sets of different size.

Contingency Table

A contingency table displays the number or frequency of observation in each of the cells formed by the intersection of different levels of two or more classification variables. Bivariate data set is another term for two measurements made on a single population. For example, information on whether men and women were smokers or non-smokers could be placed in a contingency table. There would be a frequency for men who smoke and a frequency for men who do not smoke. There would also be a frequency for women who smoke and women who do not smoke. The contingency table is most useful in displaying nominal and ordinal data.

Table 9 – Contingency Table

	Smoke	Do Not Smoke	Row Total
Males	21	17	38
Females	6	14	20
Column Total	27	31	Grand Total 58

Stem and Leaf Plot

The stem and leaf plot is a quick and easy way to display quantitative data measured on the interval or ratio scale. It is useful in arranging the observations from smallest to largest so that specific locations within the data set can be found. Each observation is divided into a stem and leaf. The first one or two digits specify the class interval and is called the *stem* and the next digit is used to construct increments which are called the *leaves*.

For example, a sample being studied had 14 participants from age 18 to age 50. The participants were ages 18, 19, 19, 19, 19, 19, 20, 20, 45, 45, 46, 47, 48, and 50. A stem and leaf plot can be created as follows.

Table 10 – Stem and Leaf Plot

```
----------------------------------------------------------------------
Stem and Leaf Display
----------------------------------------------------------------------
Field=AGE, Leaf unit= 1 , N =  15

 7     1|8999999
(2)    2|00
 6     3|
 6     4|55678
 1     5|0

The numbers to the far left are the counts of leafs
for a particular branch. The number with parentheses ()
indicates which branch contains the median of the data.
```

The stem and leaf plot makes it east to see that there were 7 participants in their teens, 2 in the 20s, none in the 30s, 5 in the 40s and 1 in the 50s. The stem and leaf can be rotated 90° clockwise and the stem will be along the left margin and the leaves on a line to the right.

Scatterplot

A Scatterplot or Scatter Diagram is a picture of the direct or inverse relationship between two variables. In addition to illustrating the direction of relationship between two variables, a Scatterplot displays the degree of association. In a Scatterplot each point represents 2 observations. The position of each point is at the intersection of the value of a single x observation and a single y observation in a Cartesian coordinate system[17]. The horizontal axis is usually known as the x-axis and the vertical axis is known as the y-axis. The points of intersection do not show the frequency of occurrences of each pair.

The bivariate data displayed in a Scatterplot is interval or ratio measurement. The Scatterplot is useful in detecting trends in the data or in revealing an association between the two variables of the bivariate data. (Kitchens 1987)

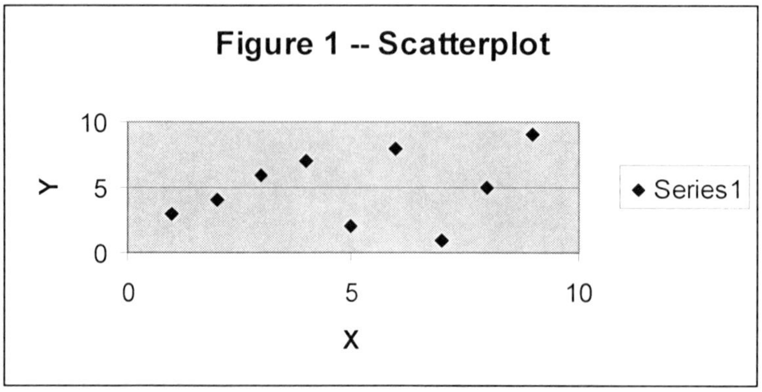

Graphs

A graph is a line connecting points representing the coordinated values of two variables.

Bar Graph

A bar graph displays the data from a frequency table. The graph consists of a horizontal and vertical axes with rectangles representing the frequency and/or relative frequency of the data. One variable is listed along the x-axis and another variable along the y-axis. Usually each bar is equal in width and separated from other bars to show that the variable being graphed on the

[17] The Cartesian coordinate system was named for the French philosopher-mathematician Descartes (1596-1650). He authored *Physica et Mathematica* in 1701.

horizontal axis is not continuous. Frequency of occurrence or magnitude of the data is shown by the height of the bar. Bar graphs can accommodate larger numbers of categories at any level of measurement. Bar graphs can also run horizontally with a vertical base line.

Figure 2 -- Bar Chart

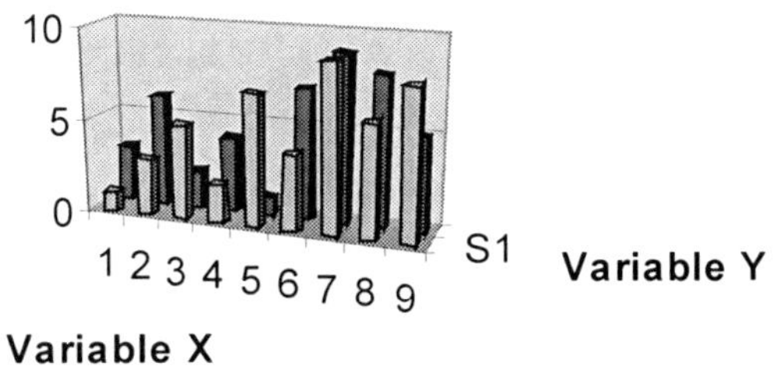

Histogram

A histogram consists of vertical bars that are connected to each other, to show that a variable measured at the interval or ratio level has the property of equal distance in its values. In constructing a histogram to represent a grouped

Figure 3 -- Histogram

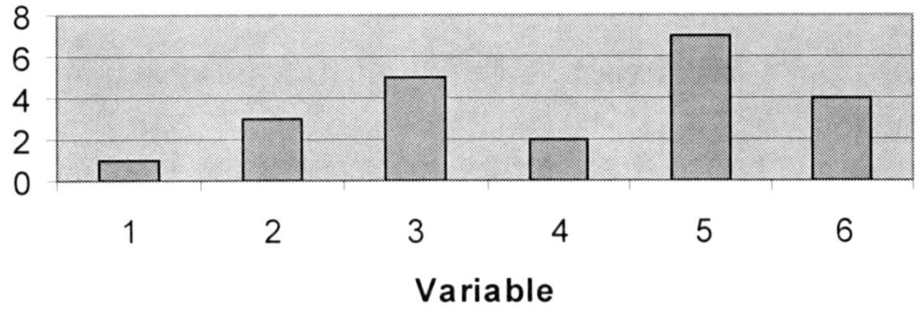

frequency distribution, the class intervals are arranged along the x-axis and the width of each bar corresponds to the width of the class interval. Rectangles or bars whose heights correspond to the respective class frequencies are drawn.

Histograms are primarily employed to plot data defined at the interval level. (Tai 1978)

Pie Chart

A pie chart is a circular graph whose pieces add up to 100 percent. A pie chart is especially useful in displaying relative frequency of categories or classes of a variable. Each category is allotted a portion of the pie equal to its relative frequency. The frequency and relative frequency as a percentage is shown inside each slice.

Figure 4 -- Pie Chart

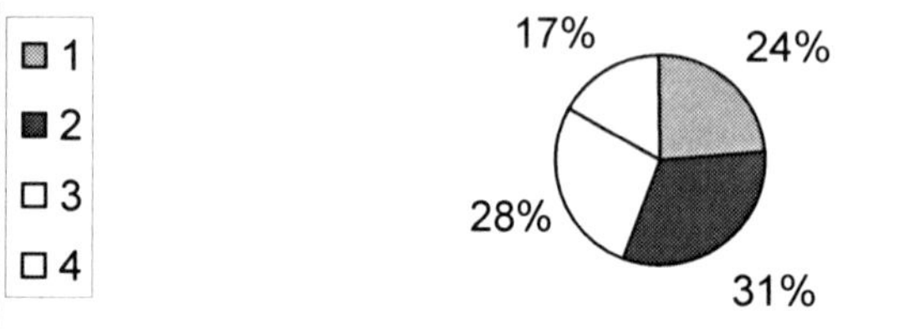

Frequency Polygon

A frequency polygon displays frequencies by indicating a series of points placed over the score values or midpoints of each class interval. The adjacent points are connected by a line, which is dropped to the base line at both ends. Frequency Polygon is similar to the Histogram except the magnitude of data is displayed by lines connecting the midpoints instead of by bars Frequency polygons are preferred for more sophisticated audiences and for graphing several distributions with equal Ns on the same chart. (Elmore et al. 1997)

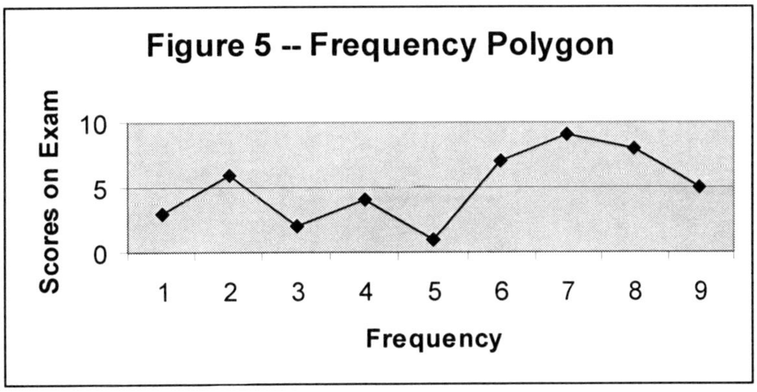

Microsoft Programs

Tables

Microsoft Office 97[18] has a number of graphic programs. Microsoft Word has three programs for developing tables. First, a Table program can be accessed from the Menu bar of the Word window. To construct a table at a position in a text, use the "Insert Table" button on the Standard toolbar. This program is highly versatile and can be used for placing text in rows and columns. A new table will consist of rows and columns of empty cells. The lines defining these cells are known as gridlines. Practically all aspects of the table can be modified through the use of the Table Menu Commands on the Menu bar. The lines which outline the tables in the beginning can be removed in the final display.

Secondly, if the table contains complex calculations, statistical analysis or charts, an Excel worksheet can be inserted. An *Enter Microsoft Excel Worksheet* button is on the standard toolbar of the Microsoft Word window.

---

[18] Halvorson and Young (1997) developed an extensive user's manual, *Running Microsoft Office 97*.

The third alternative is the *Draw Table*. To draw a table, choose *Draw Table* from the Table menu or click the *Tables and Borders* button on the standard toolbar. The *Tables and Border* toolbar has a number of options in drawing tables.

Charts

Microsoft Excel has a number of chart options. On the standard toolbar is a *Chart Wizard* button, a *Map* button, and a *Drawing* button. The *Chart Wizard* provides 14 different formats including Bar, Line, XY Scatterplot, and Pie. A first step in creating a chart is to prepare a worksheet with rows, columns, and appropriate labels. With the information in an Excel worksheet and highlighted, clicking the *Chart* command from the *Insert* menu or the *Chart Wizard* button from the standard toolbar starts the procedure for developing a chart. Options will be presented during the procedure. A Chart toolbar appears when a chart is active in the workbook and can be used in creating or modifying a chart. The *Drawing* option on the standard toolbar can be used to display a multipurpose Drawing toolbar at the bottom of the Excel window.

MINITAB

The MINITAB program has several categories of graphs and charts. The Graph button on the toolbar includes all of the basic graphs and charts mentioned above plus many others such as Time Series Plot, Matrix Plot, and Probability Plot. All are accessible from the one toolbar option.

Exercises

1. Prepare a computer generated Frequency Table. The numbers and frequencies are (1) the numbers 75-78 and the frequency is 10, (2) the number 79-82 and the frequency is 4, the numbers 83-86 and the frequency is 14, the numbers 87-90 and the frequency is 15. The table should be no more than 3 inches wide, center in the page, labeled appropriately, and single-spaced.
2. Create a computer generated 4x3 Contingency Table. The freshman, sophomore, junior, and senior class voted a preference for their favorite sport. The freshman voted 22 for football, 19 for basketball and 6 for tennis. The sophomores voted 16 for football, 26 for basketball and 9 for tennis. The juniors voted 15 for football, 23 for basketball and 12 for tennis. The seniors voted 30 for football, 15 for basketball and 15 for tennis. The table should be single spaced as far as possible, not over 4 inches wide, correctly labeled, and centered on the page.
3. Create a computer generated Scatterplot and Bar Chart. The X variable numbers are 6, 7, 5, 4, and 7. The Y variables are 78, 82, 71, 79, and 75. Label the graphics appropriately and center them on the page.
4. Create a computer generated Histogram and Pie Chart using the variables of 78, 82, 71, 79, 75, 78, 82, 71, 79, 78, 82, 71, 78, 82, and 78. Label the graphics appropriately and center them on the page.

CHAPTER 6

INDEPENDENT GROUP *t*-TEST

Basic Concept

The *t*-Test is useful with the comparison of two means, such as the comparison of a control group to an experimental group when the same type of measurement has been obtained for each group. Independent group analysis is appropriate when observations are taken from groups in which subjects in one group do not appear in another group. The observations within as well as between groups are independent of each other.

In order to apply a *t*-Test, two assumptions must first be tested and determined. The squares of the variances are assumed to be either equal[19] or not equal[20]. The *F*-Test determines if the squares of the variances are equal or not equal. The two versions of the *t*-Test (squares of the variances are equal or not equal) both have the same hypothesis and differ only in the application of the degrees of freedom. The *t*-Test with equal variances is considered the most powerful technique for the comparison of two groups measured the same way, when all necessary conditions are met.

If the population variances are not equal, the *t*-Test for unequal variances can be used. All conditions except the equality of variances should be met to use the t-Test for unequal variances appropriately.

---

19 $\sigma_1^2 = \sigma_2^2$

20 $\sigma_1^2 \neq \sigma_2^2$

Requirements

The necessary conditions to apply the t-Test include (1) two groups are independent and normally distributed, (2) the measurements are interval in nature (3) the populations from which random samples are drawn are normally distributed, (4) the population variances are equal as shown through the *F*-Test

Procedure

First, state the null hypothesis.

Second, find the critical value. The critical values for the F-Test are in the F-distribution table. The degrees of freedom are (n_1-1) across the top of the distribution and (n_2-1) down the left-hand column (See Appendix 2).

Third, determine the equality of variances with the following procedure.

$$F_V = \frac{\left[\frac{s_1^2}{\sigma_1^2}\right]}{\left[\frac{s_2^2}{\sigma_2^2}\right]} = \frac{s_1^2}{\sigma_2^2} = \frac{s_1^2}{s_2^2} = \frac{s_2^2}{s_1^2} \qquad (6.1)$$

s_1^2 = Variance of a sample

σ_2^2 = Variance of a population

Degrees of Freedom: n_1-1, and n_2-1

The *F*-scores should be above one. The larger s^2 is placed in the numerator and the smaller s^2 in the denominator.

Fourth, compare the computed *F*-value to the Tabled *F*-value (See Appendix 2, for Distribution of F). In the *F*-table read across the table to the degrees of freedom associated with the numerator. Then read down the table to the degrees of freedom associated with the denominator. Where the column and row meet, two values are stated -- a .05 value and a .01 value.

Fifth, the basic procedure for both *t*-Tests, with equal variances and without equal variances, is:

$$t = \frac{(\overline{x_1} - \overline{x_2}) - (\mu_1 - \mu_2)}{\sqrt{\frac{s_1^2}{n_1} + \frac{s_2^2}{n_2}}} \qquad (6.2)$$

$\overline{x}$ = Sample mean

μ = Population mean

n = number

Sixth, If the computed *F*-value is less than the tabled *F*-value, the variances are assumed to be equal and the degrees of freedom can be applied.

Degrees of Freedom = $n_1 + n_2 - 2$

If the computed *F*-value exceeds the tabled *F*-value the variances are assumed to be unequal and the appropriate degrees of freedom can be applied. The degrees of freedom with unequal variances are found by:

$$df = \frac{\left[\frac{s_1^2}{n_1} + \frac{s_2^2}{n^2}\right]^2}{\left[\frac{\frac{s_1^2}{n_1}}{n_1 - 1}\right]^2 + \left[\frac{\frac{s_2^2}{n^2}}{n_2 - 1}\right]^2} \qquad (6.3)$$

df = Degrees of freedom

s^2 = Variance of the sample

n = Number of the sample

Seventh, compare the calculated t-value with the table t-value. Regardless of which *t*-Test option is used, if the computed value of *t* is less than the tabled value of *t*, the hypothesis that $\mu_1 = \mu_2$ is accepted and the two samples are assumed to be from the same population. If the computed value of *t* is greater than the tabled value of *t* the hypothesis that $\mu_1 = \mu_2$ is rejected and it is assumed that the samples came from different populations.

General Example

Hungry Sam's Discount Grocery and Oscar's Save-A-Little Food Haven both have an overstock of Cajun spaghetti sauce. The price and usual amounts sold in the two stores are identical. Hungry Sam keeps the sauce in its regular place while Oscar piles the cans close to the checkout counters for a ten-day quick sale. The two storeowners record the number of cans sold during 10 days with the following results. Hungry Sam's 38, 40, 40, 42, 36, 40, 38, 44, 46, and 40; Oscar's 52, 48, 35, 33, 50, 48, 44, 52, 54, and 36.

Is there a significant difference at the 0.05 level in the number of cans sold? What is the calculated value of t? What is the table value of t?

First, state the null hypothesis. Between Sam's and Oscar's, there was no significant difference in the number of cans of Cajun spaghetti sauce sold.

Second, determine the critical values. The table value of F with (9, 9) degrees of freedom is 3.18.

Third, organize the array of numbers into a distribution and calculate the descriptive statistics, deviations, deviations squared and summed.

Table 11 – t-Test Worksheet

Hungry Sam's			Oscar's		
# Sold (x)	$(x - \bar{x})$	$(x - \bar{x})^2$	# Sold (x)	$(x - \bar{x})$	$(x - \bar{x})^2$
38	-2.4	5.76	52	6.8	46.24
40	-0.4	0.16	48	2.8	7.84
40	-0.4	0.16	35	-10.2	104.04
42	1.6	2.56	33	-12.2	148.84
36	-4.4	19.36	50	4.8	23.04
40	-0.4	0.16	48	2.8	7.84
38	-2.4	5.76	44	-1.2	1.44
44	3.6	12.96	52	6.8	46.24
46	5.6	31.36	54	8.8	77.44
40	-0.4	0.16	36	-9.2	84.64
Σ=404 $\bar{x}$=40.4		$\sum(x - \bar{x})^2 = 78.4$	Σ=452 $\bar{x}$=45.2		$\sum(x - \bar{x})^2 = 547.6$

Minitab Procedure

1. State the Research (H_a) and Null (H_o) Hypotheses in Word.
2. Open Minitab and enter the data in columns. Label the first column, Sam's; label the second column Oscar's.
3. Enter the data under the two columns.
4. Click on Stats/Basic Statistics/2-sample t-test.
5. Click on Sample in different columns.
6. Place the curser in First. Click on Sam's.
7. Place the curser in Second. Click on Oscar's.
8. Click OK
9. Copy the results from Minitab and paste them below the hypotheses in Word.
10. Summarize the results using the format below:

H_a There was a statistically significant difference in the amount of spaghetti sauce sold between Sam's and Oscar's.

H_o There was no statistically significant difference in the amount of spaghetti sauce sold between Sam's and Oscar's.

Two-Sample T-Test and CI: Sams, Oscars

```
Two-sample T for Sams vs Oscars

         N   Mean  StDev  SE Mean
Sams    10  40.40   2.95     0.93
Oscars  10  45.20   7.80      2.5
```

```
Difference = mu (Sams) - mu (Oscars)
Estimate for difference:  -4.80
95% CI for difference:  (-10.60, 1.00)
T-Test of difference = 0 (vs not =): T-Value = -1.82  P-Value = 0.096  DF = 11
```

Since the level of significance was greater than 0.05 (p=0.096), the null hypothesis was accepted and the research hypothesis was not supported. There was no difference in the amount of spaghetti sauce sold between Sam's and Oscar's.

Exercise

The Wonderful Worshipful Order of Wolverine Watchers raised funds for the preservation of the wolverine. There was an ongoing dispute about the value of identifying donors. Those that disagreed agreed to raise funds the first year with no public identification of donors. The second year, they agreed to place a full-page ad in the Upper Peninsula World Guardian listing all of the donors to the Wolverine Preservation Fund. Following are representative funds for each year.

Table 12
Wolverine Protection Fund Raising Worksheet

First Year Donor Anonymity	Second Year Donor Identity
4	3
3	5
5	5
6	5
2	4
3	5

Use Minitab to determine the *p* value of the *t*-Test following the format above to report the hypotheses, results, and conclusion.

CHAPTER 7

ONE-WAY ANOVA[21]

Basic Concept

When the means of three or more independent groups are compared, one-way or single factor analysis of variance techniques are appropriate. Although comparing three or more means, the One-Way Analysis of Variance compares variances instead of means. Treatment of the data using the analysis of variance requires the F_a-Test. With the F test two different estimates of the population variance are made – between-group variance and within-group variance.

Requirements

The conditions under which the F_a-test can be applied are as follows: (1) The three or more groups must be independent. (2) The measurements must be interval in nature. (3) The populations must be normally distributed. (4) The population variances must be equal as shown by an F-Test. (5) The samples must be randomly drawn.

Procedure

Calculating an ANOVA is a process that involves at least three formulas and two tables.

First, state the null hypotheses.

Second, determine the critical value. (See Appendix 2)

[21] ANOVA is an abbreviation of Analysis of Variance

Third, collect information from the samples. This information may be prepared as worksheet 1.

Table 13 – ANOVA Worksheet 1

Unknown	Sample 1	Sample 2	Sample 3	Sample k	Total
ΣX	ΣX_1	ΣX_2	ΣX_3	ΣX_k	ΣX
ΣX^2	ΣX_1^2	ΣX_2^2	ΣX_3^2	ΣX_k^2	ΣX^2
n	n_1	n_2	n_3	n_k	N

Fourth, find the between-group and within-group sum of squares.

$$Between(SS_b) = \left[\frac{(\Sigma x_1)^2}{n_1} + \frac{(\Sigma x_2)^2}{n_2} + \ldots + \frac{(\Sigma x_k)^2}{n_k}\right] - \frac{(\Sigma x)^2}{N} \quad (7.1)$$

$$Within(SS_w) = \left[\Sigma x_1^2 - \frac{(\Sigma x_1)^2}{n_1}\right] + \left[\Sigma x_2^2 - \frac{(\Sigma x_2)^2}{n_2}\right] + \ldots + \left[\Sigma x_k^2 - \frac{(\Sigma x_k)^2}{n_k}\right] \quad (7.2)$$

$$Total(SS_T) = \sum x^2 - \frac{\left(\sum x\right)^2}{N} \quad (7.3)$$

Fifth, prepare ANOVA Worksheet 2.

Table 14 – ANOVA Worksheet 2

Source of Variation	SS	Degrees of Freedom	Mean Square (MS)	Computed F_a
Between	SS_b	k - 1	$\frac{SS_b}{df_b}$	$\frac{MS_b}{MS_w}$
Within	SS_w	N – k	$\frac{SS_w}{df_w}$	
Total	SS_T	N -1		

Sixth, apply the F_A procedure.

$$F_a = \frac{MS_b}{MS_w} \quad (7.4)$$

Seventh, compare the calculated *F*-test value with the table value.

An accepted null hypothesis brings to an end the statistical analysis of a study. There is no significant difference among the groups. However, a rejection of the hypothesis of no difference results in the need to identify where the significant differences lie.

Eighth, to find the groups in which the difference lies, a number of statistical procedures are available including Duncan, Tukey, Newman-Keuls or Scheffe.[22] The Minitab program provides three options: Newman-Keuls, Scheffe, and Tukey. Different techniques are used among the Minitab options causing the results to vary. The Newman-Keuls and Tukey tests are considered by some to be the most commonly used multiple comparison procedures. Dunnett's test is a specialized multiple comparison test that allows comparison of a single control group to all other groups.

Tukey Honestly Significant Difference Test

To perform the Tukey Honestly Significant Difference (HSD) test the following formula is applied to each pair of means.

$$q = \frac{\bar{x}_j - \bar{x}_k}{\sqrt{\frac{MS_w}{n}}} \qquad (7.5)$$

q = Studentized Range Statistic

MS_w = mean square for within groups from ANOVA summary table

n = number of observations per subgroup

[22] For a more complete discussion of the Tukey procedure and the Duncan multiple range test see Freund et al. (1997), pages 247-256. See Elmore et al. (1997) pages 198-203 for information about Multiple Comparison Procedures including the Tukey HSD test, the Newman-Keuls procedure and the Scheffe procedure. See Levin (1983) pages 165-167 for a discussion of the Tukey procedure. See the topic Multiple Comparisons in Hinton (1995) pages 130-139 for a discussion of the Tukey test and the Scheffe test.

A convenient way to run the Tukey HSD procedure is to create tables of mean differences and corresponding q-values.

General Example

The following scenario illustrates the use of an ANOVA to analyze means for significant difference. St. Louis players totaled 2, 16, 28, 2, 21, 2, 2, 19, 0, 4, 8, and 6 home runs. Atlanta players hit 9, 2, 1, 10, 13, 1, 12, 0, 1, 2, 7, and 7 home runs. Jacksonville players hit 13, 39, 25, 9, 3, 0, 13, 27, 5, 1, and 2 home runs. Test the hypothesis that there is no significant difference in the total number of home runs hit by the three ball clubs.

First, state the null hypothesis. There is no significant difference in the total number of home runs hit by the three ball clubs.

Second, find the critical value. The degrees of freedom for the numerator

Table 15 -- ANOVA Worksheet 1
Home Runs Analysis

St. Louis X_1	X_1^2	Atlanta X_2	X_2^2	Jackville X_3	X_3^2	Total
2	4	9	81	13	169	
16	256	2	4	39	1521	
28	784	1	1	25	625	
2	4	10	100	9	81	
21	441	13	169	3	9	
2	4	1	1	0	0	
2	4	12	144	13	169	$\Sigma X=312$
19	361	0	0	27	729	
0	0	1	1	5	25	
4	16	2	4	1	1	
8	64	7	49	2	4	$(\Sigma X)^2=$ 97,344
6	36	7	49			
$\Sigma x_1=110$		$\Sigma x_2=65$		$\Sigma x_3=137$		
	$\Sigma x_1^2=$ 1974		$\Sigma x_2^2=$ 603		$\Sigma x_3^2=$ 3333	$\Sigma X^2=$ 5910
n=12		n=12		n=11		N=35

is k – 1 or 3 – 1 = 2; the degrees of freedom for the denominator is N – k or 35 – 3 = 32. The tabled value of F with 2 degrees of freedom in the numerator and 32 degrees in the denominator is between 3.32 (30 df) and 3.27 (35 df). (See Appendix 2, page 167)

Third, using the information in Worksheet I calculate the sum of the squares between and within three sets of numbers.

$$Between(SS_b) = \left[\frac{(\Sigma x_1)^2}{n_1} + \frac{(\Sigma x_2)^2}{n_2} + \ldots + \frac{(\Sigma x_k)^2}{n_k}\right] - \frac{(\Sigma x)^2}{N} \quad (7.1)$$

$$Between(SS_b) = \left[\frac{(12{,}100)}{12} + \frac{(4225)}{12} + \frac{(18{,}769)}{11}\right] - \frac{(97{,}344)}{35}$$

$$= [1008.33 + 352.08 + 1706.27] - 2781.26$$

$$= 285.42$$

$$Within(SS_w) = \left[\Sigma x_1^2 - \frac{(\Sigma x_1)^2}{n_1}\right] + \left[\Sigma x_2^2 - \frac{(\Sigma x_2)^2}{n_2}\right] + \ldots + \left[\Sigma x_k^2 - \frac{(\Sigma x_k)^2}{n_k}\right] \quad (7.2)$$

$$Within(SS_w) = \left[1974 - \frac{(110)^2}{12}\right] + \left[603 - \frac{(65)^2}{12}\right] + \left[3333 - \frac{(137)^2}{11}\right]$$

$$= [1974 - 1008.33 + 603 - 352.08 + 3333 - 1706.27]$$

$$= 2843$$

Fourth, test the accuracy of the calculation of the within and between sum of the squares with the following procedure. SS_b and SS_w are added and the sum is the total sum of squares, SS_T.

$$SS_b + SS_w = SS_T \quad (7.3)$$

285.42 + 2843 = 3128.74

When the calculation of the sum of the squares is accurate, the sum of the squares is equal to the number produced by the following formula.

$$Total(SS_T) = \Sigma X^2 - \frac{(\Sigma X)^2}{N} \tag{7.3}$$

$$= [1974 + 603 + 3333] - \frac{(312)^2}{35}$$

= 5910-2781.26 = 3128.74

Fifth, prepare ANOVA Worksheet 2. Worksheet 2 provides an organization of the sum of squares, degrees of freedom from which the mean square is calculated. (See Table 16)

Table 16-- ANOVA Worksheet 2
Home Runs Analysis

Source of Variation	Sum of the Squares (SS)	Degrees of Freedom (df)	Mean of Squares (MS)	Computed F_A
Between	SS_b	k -1	$\frac{SS_b}{df_b}$	$\frac{MS_b}{MS_w}$
Within	SS_w	N - k	$\frac{SS_w}{df_w}$	
Total	SS_T	N - 1		

Sixth, calculate the analysis of variance model or ANOVA by dividing the mean sum of squares by the within mean sum of squares.

$$F_A = \frac{MS_b}{MS_w} \tag{7.4}$$

$$F_A = \frac{142.5}{88.84} = 1.604$$

Seventh, compare the computed F_a to the tabled F_a. The degrees of freedom associated with the numerator (MS_b) are read across the table and the degrees of freedom associated with the denominator (MS_w) are read down the side of the table. If the computed F_a is larger than the tabled F_a the hypothesis is rejected. If the computed F_a is smaller than the tabled F_a the hypothesis is accepted.

The tabled value of F with 2 degrees of freedom in the numerator and 32 degrees in the denominator is between 3.32 (30 df) and 3.27 (35 df). In this instance, the hypothesis of no difference is accepted. (See Appendix 2, page 146)

Eighth, perform the Tukey Honestly Significant Difference (HSD) test by apply the following procedure to each pair of means.

$$q = \frac{\overline{x_j} - \overline{x_k}}{\sqrt{\frac{MS_w}{n}}} \tag{7.5}$$

q = Studentized Range Statistic

MS_w = mean square for within groups from ANOVA summary table

n = number of observations per subgroup

A convenient way to run the Tukey HSD procedure is to create tables of mean differences and corresponding q-values.

Table 17 -- Tukey HSD Test Worksheet
Home Run Analysis

Means Being Compared	Mean Difference	q-value
Mean 3 & 2	7.0379	2.53
Mean 3 & 1	3.2879	1.18
Mean 1 & 2	3.75	1.35

Following is the procedure for calculating the q value in the comparison of group 3 and 2 or the comparison between the number of home runs for Atlanta and Jacksonville for a season.

$$q = \frac{12.4545 - 5.4167}{\sqrt{\frac{88.85}{35}}} \qquad (7.5)$$

$$= \frac{7.0379}{\sqrt{\frac{88.85}{35}}} = \frac{7.0379}{2.7766} = 2.5347$$

The tabled value of q with 32 df = 3.477 (See Appendix C). The number of means is 3 and the Error degree of freedom is 32. The degrees of freedom within is the total number of observations minus the number of columns. The procedure above is used to calculate the q value for each set of groups. (See Appendix 9, page 162)

Since the calculated value of q is larger than the tabled value of q, there is no significant difference between any of the three groups. If the table value of q had been larger than the calculated value, there would be significant difference between the two groups being tested.

Minitab Procedure

1. State the research (H_a) and the null (H_o) hypotheses in Word.
2. Create a database by opening Minitab and labeling three columns: St. Louis, Atlanta, and Jacksonville. Then enter the data in the columns.
3. Analyze the data by clicking on Stat, ANOVA, and One Way (Unstacked). In the ANOVA window identify the Responses by double-clicking on each column. Click on Comparisons and select Tukey. Click OK and OK to get the results.

4. Copy the Minitab results and paste them below the hypotheses in Word.
5. Summarize the results as below:

H_a There was a statistically significant difference in home runs among the three ball teams.

H_o There was no statistically significant difference in home runs among the three ball teams.

One-way ANOVA: St. Louis, Atlanta, Jacksonville

```
Source  DF      SS     MS     F      P
Factor   2   285.4  142.7  1.61  0.216
Error   32  2843.3   88.9
Total   34  3128.7

S = 9.426   R-Sq = 9.12%   R-Sq(adj) = 3.44%

Level          N    Mean   StDev
St. Louis     12   9.167   9.370
Atlanta       12   5.417   4.776
Jacksonville  11  12.455  12.754

              Individual 95% CIs For Mean Based on Pooled StDev
Level           +---------+---------+---------+---------
St. Louis              (----------*----------)
Atlanta         (----------*----------)
Jacksonville                 (-----------*----------)
                +---------+---------+---------+---------
              0.0       5.0      10.0      15.0

Pooled StDev = 9.426
```

Since the level of significance was greater than 0.05 (p=0.216), the null hypothesis was accepted and the research hypothesis was not supported. There was no statistically significant difference in home runs among the three ball teams.

Exercises

1. Three types of computer disks are selected and the number of defects in each is as follows. Type A has 0, 1, 0, 2, 3, 2, 0, 1, 1, and 0. Type B has 2, 0, 3, 5, 3, 4, 6, 0, 2, and 5. Type C has 1, 0, 1, 1, 0, 2, 0, 0, 1, and 2. Assume a normal distribution and use ANOVA to analyze the defects. Report the hypotheses, Minitab printout, and conclusion in the format used for the general example above.
2. Analyze the difference between disk type B and disk type C above using the appropriate statistical analysis (use the 4 questions and the Decision Model). Report your findings using the format for hypotheses, Minitab printout, and conclusion as in the general example above.

CHAPTER 8

PAIRED t-TEST

A Paired *t*-Test for Repeated Measures compares two measurements from one group. R.A. Fisher[23] developed several formulas[24] designed to compute *t*. The Independent Group *t*-Test was a test for difference between two independent sample means of equal size. Another application of the t-Test is the Paired t-Test, which compares two means that are paired or matched in some way. The test corresponds to the Minitab Basic Statistic of Paired t.

Requirements

Five conditions must exist to employ the Paired *t*-Test correctly for difference between two related means from samples of equal size. (1) A single group must contribute the two sets of data or two groups must be deemed one. (2) Measurements must be at the interval level. (3) The population must be normally distributed. (4) The population variances must be equal as shown through an *F* test. (5) Samples must be randomly drawn.

Procedure

Following is the procedure for the Fisher *t* for testing the difference between uncorrelated means in two samples of equal size.

First, state the null hypothesis.

Second, find the critical value in the t-distribution. The degrees of freedom = N – 1, where N is the number of pairs. (See Appendix 1, page XXX)

---

[23] Sir Ronald Fisher (1890-1963) was a renowned English statistician. (Kitchen 1987)

[24] Fisher developed a *t* ratio for testing the significance of a coefficient of correlation. He developed a second *t* for testing a difference between uncorrelated means. He developed a third *t* to test for difference between uncorrelated means in two samples of equal size. Fisher developed a fourth *t* to test the differences between correlated pairs of means. (Guilford 1965)

Third, develop a table with the first measurement, the second measurement, the difference between the two measurements, and the square of the differences. Calculate the mean for the first measurement, the mean for the second measurement, the sum of the differences, and the sum of the differences squared.

Fourth, apply the procedure to calculate the value of t.

$$t = \frac{M_1 - M_2}{\sqrt{\frac{N\left(\sum D^2\right) - \left(\sum D\right)^2}{N^2(N-1)}}} \tag{8.1}$$

M_1 = The Mean of the first measurement

M_2 = The Mean of the second measurement

D = The Difference between the first measurement and the second Measurement

N = The Number of Pairs

The degrees of freedom in this procedure is N – 1, where N is the number of pairs.

Fifth, compare the calculated value of t with the table value of t. If the calculated value of t is greater than the table value of t, the null hypothesis of no difference is rejected. If the calculated value of t is less than the table value of t, the null hypothesis of no difference is accepted.

General Example

Alister Armbruester's Alligator Alley Fat Farm and School of Ballet advertised guaranteed weight loss in two weeks on his special alligator meat and seaweed diet. When the local Better Business Bureau challenged the guarantee, Alister promptly demonstrated the weight loss by means of a random sample of

10 men. The 10 men's weight was measured the day they enrolled in the special alligator meat and seaweed diet and measured again after two weeks. The beginning and ending measurements were 182 / 177, 173 / 173, 191 / 184, 188 / 189, 213 / 200, 203 / 200, 167 / 159, 174 / 167, 184 / 179, and 185 / 178.

First, state the null hypothesis. Between the beginning weights and the ending weights, there is no significant difference.

Second, find the critical difference. The table value for a one tail test at the 0.05 level with 9 degrees of freedom, N-1, was +1.833. [One tail because the hypothesis requires loss of weight.] (See Appendix 1, page 144)

Third, develop a table with weight before, weight after, the difference between the before and after, and the square of the differences. Calculate the mean for the before weights, the mean for the after weights, the sum of the differences, and the sum of the differences squared.

Table 18 -- Paired t-Test Worksheet
Weight Loss Analysis

Weight Before	Weight After Diet	Difference (D)	Difference Squared (D^2)
182	177	5	25
173	173	0	0
191	184	7	49
188	189	-1	1
213	200	13	169
203	200	3	9
167	159	8	64
174	167	7	49
184	179	5	25
185	178	7	49
$M_1 = 186$	$M_2 = 180.6$	$\Sigma D = 54$	$\Sigma D^2 = 440$

Fourth, apply the procedure to determine the value of t.

$$t = \frac{M_1 - M_2}{\sqrt{\frac{N\left(\sum D^2\right) - \left(\sum D\right)^2}{N^2(N-1)}}} \quad (8.1)$$

$$t = \frac{186 - 180.6}{\sqrt{\frac{10(440) - (54)^2}{100(9)}}} = \frac{5.4}{\sqrt{\frac{1484}{900}}} = \frac{5.4}{\sqrt{1.649}} = \frac{5.4}{1.284} = 4.205$$

The calculated Fisher's t for testing the difference between two measurements of a single sample, Paired t-Test, was 4.205.

Fifth, make a decision. Compare the calculated value with the table value. The table value at the 0.05 level with 9 degrees of freedom, N-1, was +1.833. Since the computed t value was larger than the table value, the null hypothesis of no difference in weight before and after the two-week diet was rejected. The research hypothesis of a significant difference after the diet was supported. The Better Business Bureau admitted that the alligator meat and seaweed diet worked as advertised.

Minitab Example

1. State the research hypothesis (H_a) and null hypothesis (H_o) in Word.
2. Open Minitab and label two columns as Before Diet and After Diet.
3. Enter the values in the database.
4. Analyze the data by clicking on Stat/Basic Statistics/Paired t.
5. Put "Before Diet" in the First Sample box and "After Diet" in the Second Sample Box.
6. Click OK.
7. Copy the results from Minitab and paste them below the hypotheses in Word.
8. Summarize the results using the format below:

H_a There was a statistically significant difference in weight before and after the diet.
H_o There was no statistically significant difference in weight before and after the diet.

Paired T-Test and CI: Before Diet, After Diet

```
Paired T for Before Diet - After Diet

              N    Mean  StDev  SE Mean
Before Diet  10  186.00  13.91     4.40
After Diet   10  180.60  13.19     4.17
Difference   10    5.40   4.06     1.28


95% CI for mean difference: (2.50, 8.30)
T-Test of mean difference = 0 (vs not = 0): T-Value = 4.21  P-Value = 0.002
```

Since the level of significance was less than 0.05 (p=0.002), the null hypothesis was rejected and the research hypothesis was supported. There was a statistically significant difference in weight before and after the diet.

Exercises

1. A coach claims that a weight lifter can increase his strength by taking a special vitamin. Eight athletes are selected and given a test of strength, using the standard bench press. After two weeks of regular training supplemented with the vitamin, they are tested again. Test the effectiveness of the vitamin regimen at $\alpha = 0.05$. Following are the before and after weights that each of the athletes bench pressed: 210 / 219, 230 / 236, 182 / 179, 205 / 204, 262 / 270, 253 / 250, 219 / 222, and 216 / 216. Assume that the variable is approximately normally distributed. Report the hypotheses, Minitab printout, and conclusion using the same format as the general example above.

2. Assume that the first weight for each lifter was reported for one group of athletes and the second weight for each lifter was reported for a second group of athletes. Assume that the weight variable is approximately normally distributed. Using the four questions and the Decision Model, select the proper statistic and

report the hypotheses, Minitab printout, and conclusion in the same format as above in the general example. (Hint—the data are Independent)

CHAPTER 9

PEARSON PRODUCT MOMENT CORRELATION

Basic Concept

Correlation is a statistical method of determining whether a relationship exists between two variables. Regression is a statistical method that describes the nature of the relationship between variables – whether positive or negative, linear or nonlinear.

The Pearson Product Moment Correlation determines whether or not two measurements taken on a single group produce a significant relationship between the two measurements. However, finding that a relationship exists may not indicate much about the degree of association or correlation between two variables. Many relationships are statistically significant; few express perfect or exact correlation. In correlation research, it usually is assumed that the pairs of values are sampled from a bivariate normal distribution, meaning that for any selected value of one variable, the values of the other variable are normally distributed.

Requirements

Five conditions must exist to employ the Pearson Product Moment of Correlation as a measure of association between two variables. (1) A single group must contribute the two sets of data or two groups must be deemed one. (2) Measurements must be interval. (3) The population must be normally distributed. (4) The population variances must be equal as shown through an *F* test. (5) Samples must be randomly drawn.

Correlation Coefficient

With the aid of the Pearson correlation coefficient *r* the strength and direction of relationship between two variables can be determined. The correlation coefficient is defined as the covariance of the variable divided by the product of the standard deviations of the variables.

The measurement of the relationship is a mathematical value between –1 and +1. Values close to +1 indicate that the measurements achieved by the group on a first measure were nearly identical to the measurements on the second measure. Values that are close to –1 indicate that the measurements achieved by the group on the first measure were the reverse of the measurements on the second measure. Values near 0 indicate that neither of the two conditions described above predominated. Some subjects follow the pattern of the +1 condition, while other subjects follow the pattern of the –1 condition. Neither condition prevails in a dominant way.

Several procedures are available for calculating the Pearson r depending upon the form of the data. Pearson originally devised the formula for z-scores or a definitional formula. Covariance, deviation score and raw-score formulas are also available[25].

t-Test

The significance of the Pearson *r* is tested by computing a *t* ratio with the degrees of freedom equal to N-2 and N equal to the number of pairs of scores.

---

[25] See Elmore et al. (1997), pages 96-101 for a complete discussion of the four formulas available to calculate the Pearson r.

Procedure

Determine if there is a relationship between two measurements.

First, state the null hypothesis.

Second, determine the critical value. (See Appendix A, t Distribution)

Third, create a table with seven columns and as many rows as pairs of observations plus two additional rows. List the two sets of numbers in columns 1 and 2 and label the columns X and Y. For each set of numbers, calculate the sums and the means. In columns 3 and 4, labeled x and y, subtract the mean of X and Y from each entry in columns 1 and 2. In columns 5 and 6, labeled x^2 and y^2, calculate the square of each number in the x and y columns. In the bottom row of columns 5 and 6, total the x^2 column to obtain Σx^2 and the y^2 column to obtain Σy^2. In column 7, multiply each row in x times y to obtain xy. In the bottom row of column 7, total the xy column to obtain Σxy. Fourth, apply the deviation score procedure to determine the Pearson Product Moment Coefficient *r.*

$$r_{xy} = \frac{\sum xy}{\sqrt{\left(\sum x^2\right)\left(\sum y^2\right)}} \qquad (9.1)$$

Fifth, test the hypothesis r = 0. The test of the hypothesis is made with a Fisher t-Test.

$$t = \frac{r_{xy}\sqrt{N-2}}{\sqrt{1-\left(r_{xy}\right)^2}} \qquad (9.2)$$

Sixth, compare the computed value of *t* with the table value of *t*. If computed score is larger, the null hypothesis is rejected and the research hypothesis is supported. If the table score is larger, the null hypothesis cannot be rejected nor can the alternative hypothesis be supported.

General Example

A statistics instructor is interested in determining if mid-term exam grades and final exam grades are consistent. To determine consistence, the instructor tested a small sample to see if there was correlation between the two sets of grades. The first student scored 12 and 10, the second student 14 and 14, the third student 8 and 10, the fourth student 14 and 12, and the fifth student 12 and 14.

First, state the null hypothesis. There is no significant correlation between the mid term exam scores and the final exam scores.

Second, determine the critical value for the t-Test. The degrees of freedom is N – 2 or 5 – 2 = 3. In the Fisher's t –Table at the 0.05 level of probability, the critical value was 2.353. (See Appendix 1, t Distribution)

Third, create a table with seven columns and as many rows as pairs of observations plus two additional rows. List the two sets of numbers in columns 1 and 2 and label the columns of numbers X and Y. For each set of numbers, calculate the sums and the means. In columns 3 and 4, labeled x and y, subtract the mean of X and Y from each entry in columns 1 and 2. In columns 5 and 6, labeled x^2 and y^2, calculate the square of each number in the x and y columns. In the bottom row of columns 5 and 6, total the x^2 column to obtain Σx^2 and the y^2 column to obtain Σy^2. In column 7, multiply each row in x times y to obtain xy. In the bottom row of column 7, total the xy column to obtain Σxy. (See Table 22)

Table 19 -- Correlation Analysis Worksheet

Test A (X)	Test B (Y)	$(X - \bar{x})$ x	$(Y - \bar{y})$ y	$(X - \bar{x})^2$ x^2	$(Y - \bar{y})^2$ y^2	xy
12	10	0	-2	0	4	0
14	14	2	2	4	4	4
8	10	-4	-2	16	4	8
14	12	2	0	4	0	0
12	14	0	2	0	4	0
$\bar{x} = 12$	$\bar{x} = 12$			$\sum x^2 = 24$	$\sum y^2 = 16$	$\sum xy = 12$

Fourth, apply the procedure to determine the Pearson Product Moment Coefficient r.

$$r_{xy} = \frac{\sum xy}{\sqrt{\left[\left(\sum x^2\right)\left(\sum y^2\right)\right]}} = \frac{12}{\sqrt{(24)(16)}} = \frac{12}{\sqrt{384}} = \frac{12}{19.6} = .61 \qquad (9.1)$$

Fifth, test the hypothesis r = 0 with a Fisher t-Test.

$$t = \frac{r_{xy}\sqrt{N-2}}{\sqrt{1-\left(r_{xy}\right)^2}} = \frac{.61\sqrt{5-2}}{\sqrt{1-(.61)^2}} = \frac{.61\sqrt{3}}{\sqrt{1-.372}} = \frac{(.61)(1.732)}{\sqrt{.628}} = \frac{1.057}{.792} = 1.335 \qquad (9.2)$$

Sixth, make a decision. The calculated value of t was 1.335, which was less than the table value of 2.353 The correlation of .61 did not show significant relationship. The null hypothesis of no significant correlation was accepted. The correlation was not significant.

Scatter Plot

A scatter plot is a visual way of describing the nature of the relationship between two variables. The horizontal axis is x and the vertical axis is the

y axis.

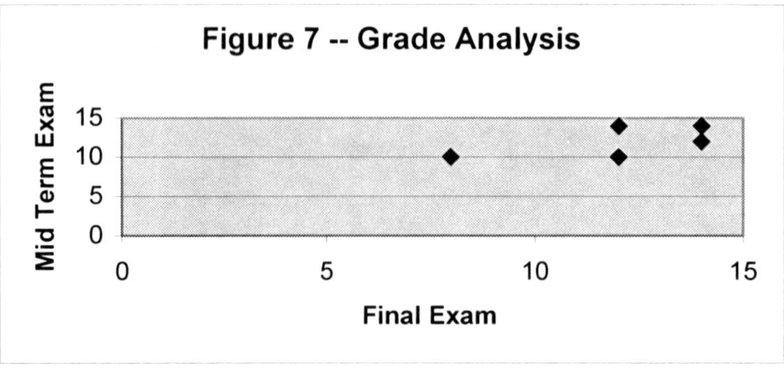

x axis

Altered Example

If the example above had 32 pairs of scores instead of 5, the outcome would have been somewhat different. With 32 pairs, the Fisher table was consulted at the .05 level of probability for 30 degrees of freedom (N – 2 or 32 – 2 = 30), the value was 2.042. The computed value of t was 4.291, which was greater than the .05 level of probability, 2.042. The correlation of .61 then showed a real or significant relationship, and not a chance relationship, since the probability was only 5 in 100 that the relationship was due to chance. The null hypothesis of no relationship would have been rejected.

MINITAB Procedure

1. State the research hypothesis (H_a) and null hypothesis (H_o) in Word.
2. Open Minitab and label two columns as Pre-Test (A) and Post-Test (B).
3. Enter the values in the database.
4. Analyze the data by clicking on Stat/Basic Statistics/Correlation.
5. Enter the variables, Pre-Test and Post-Test.
6. Make sure the Display P-Values is checked.
7. Click OK.
8. Copy the results from Minitab and paste them below the hypotheses in Word.
9. Summarize the results using the format below:

H_a There was a statistically significant correlation between the scores on test A and the scores on test B.

H_0 There was no statistically significant correlation between the scores on test A and the scores on test B.

Correlations: Pre-Test A, Post-Test B

```
Pearson correlation of Pre-Test A and Post-Test B = 0.612
P-Value = 0.272
```

Since the level of significance was greater than 0.05 (p=0.272), the null hypothesis was accepted and the research hypothesis was not supported. There was no statistically significant correlation between the scores on test A and the scores on test B.

Exercises

1. Determine the correlation coefficient *r* and if there is significant correlation between the absences and the final grades of a sample of 7 participants in a statistics class. The absences / grades for the 7 students were 6 / 82, 2 / 86, 15 / 43, 9 / 74, 12 / 58, 5 / 90, and 8 / 78. Assume that the viable are normally distributed for the sake of using the Pearson Correlation in this exercise. If the coefficient *r* should happen to be a negative number, be sure to note that the resulting correlation is an inverse correlation (as one variable increases the other decreases). Report the hypotheses, Minitab printout, and conclusion using the format for the general example above.

2. Data were collected for a second and third group of students as well. The three groups data are recorded in the table below.

Absents Group A	Grades Group A	Absents Group B	Grades Group B	Absents Group C	Grades Group C
6	82	3	79	8	86
2	86	1	83	4	90
15	43	12	40	17	47

9	74	6	71	11	78
12	58	9	55	14	62
5	90	2	87	7	90
8	78	5	75	10	82

Test for a difference in grades among the three groups of students using the appropriate statistic determined by answering the four questions and using the Decision Model. Report the hypotheses, Minitab printout, and conclusion using the format for the general example above.

3. Using the data in the table for number two above, test for a statistically significant difference in absences between Group A and Group C. Report the hypotheses, Minitab printout, and conclusion using the format for the general example above.

CHAPTER 10

MANN-WHITNEY U TEST

Basic Concept

The Mann-Whitney U Test, symbol z_u, is one of the most powerful nonparametric tests. It is applied to measure two groups when the parametric Independent Group t-Test conditions are not met. The data is based upon a small sample or a larger sample known to be from a non-normal population and/or from a population in which the variances are unequal. Mann-Whitney is also useful if exact data values are unavailable but the order of data is known.

The Mann-Whitney procedure differs from parametric measurements in that the ranks of the data are used for the analysis rather than the raw values. The test is often referred to as the U-Test, since a statistic designated as *U* is computed for testing the null hypothesis.

Requirements

The conditions necessary for applying the Mann Whitney U Test are (1) two groups are being compared, (2) samples are drawn independently, (3) the data are ordinal in nature, and (4) the data are changed from scores to ranks.

Procedure

The following is the procedure for calculating the Mann-Whitney U Test:

First, state the null hypothesis.

Second, determine α and df in order to establish the critical value.

Third, convert the data to ranks. Place the sets of numbers being compared on a worksheet in two columns with the largest values in both columns at the top of the sheet of paper. Add a second column beside each of the original columns. If the original columns are identified as X_1 and X_2, the new columns can be identified as R_1 and R_2. Find the total numbers, *N*, by adding n_1 *and* n_2. Arrange the data from the combined samples in order of magnitude and rank, with the lowest value given rank 1. Find the second lowest score and place a 2 in the appropriate R column.

When a particular score occurs more than once, assign a rank for each occurrence, add the ranks together and divide by the number of tied numbers. Place the resulting value beside each of the tied scores. The next score will be assigned the rank following the last one used in determining the value for the tied scores. The last rank assigned should be or should account for the value earlier determine as N.

Fourth, sum each of the R columns to find ΣR_1 and ΣR_2. When the resulting values of ΣR_1 and ΣR_2 are added together they should be equal to $N(N + 1)$ divided by 2.

$$\sum R_1 + \sum R_2 = \frac{N(N+1)}{2} \qquad (10.1)$$

Fifth, solve for U using the following procedure.

$$U_1 = n_1 n_2 + \frac{n_1(n_1+1)}{2} = \sum R_1 \qquad (10.2)$$

$$U_2 = n_1 n_2 + \frac{n_1(n_1+1)}{2} = \sum R_2$$

The two formulas given above will most likely result in two different values for U values. The value selected for U in hypothesis testing is the lesser of the two. To check whether the computation of the U value is correct, the following formula is used.

$$U = n_1 n_2 \text{ - the larger value of } U$$

Sixth, compare the values of U_1 and U_2 chose the smaller of the two to use in the z_u procedure.

Seventh, determine the significance of an observed value of U by applying the following formula.

$$z_u = \frac{U + \frac{1}{2} - \frac{n_1 n_2}{2}}{\sqrt{\frac{(n_1)(n_2)(n_1 + n_2 + 1)}{12}}} \qquad (10.3)$$

Eighth, make a decision. Compare the calculated z_u value to the critical value of plus or minus 0 to 1.645. If the computed z_u is larger than a plus or minus 1.645, the hypothesis of no difference is rejected. If the computed z_u is between a plus or minus 0 and 1.645, the hypothesis of no difference is accepted.

General Example

Bingo Zumbroski, dean of Bargain Basement Law School and Storm Door Company compiled data on alumni who received a law degree 5 years earlier. He divided the group into those who chose private practice and those who chose corporate law. He received data from 8 in private practice and 12 in corporate law. Four, who were installing storm doors, were not counted. Zumbroski hypothesized that the lawyers in private practice earned higher annual incomes.

First, state the null hypothesis. Between the lawyers in private practice and corporate practice there was no significant difference in income.

Second, determine the critical value when $\alpha = 0.05$. The critical value is a plus or minus 1.645.

Third, prepare a table with 6 columns. In column 1 list the names of the private practice lawyers. In column 2 list the income of the private practice lawyers. In column 3 list the overall rank, including the corporate lawyers, of the private practice lawyers. In columns 4, 5, and 6 list the same information for the corporate lawyers. Sum the ranks for the private practice lawyers and sum the ranks for the corporate lawyers. Calculate n for each of the groups.

Table 20 -- Mann-Whitney Worksheet
Lawyer Salary Analysis

Private Practice	Annual Income $ Thousands	Income Rank	Corporate Law	Annual Income $ Thousands	Income Rank
R. Hayes	22.4	15	A. Award	21.9	14
A.Newman	17.8	3	C. Music	16.8	1
D. Duck	26.5	16	H. Truman	28.0	17
W. Beery	19.3	8	B. Ruth	19.5	10
J. Louis	18.2	5.5	R. Finger	18.2	5.5
J. Doe	21.1	13	J. Blow	17.9	4
M. Free	19.7	11	J. Bugg	35.8	19
I. Nutz	43.5	20	E. Fudd	20.5	12
			B. Wheel	18.7	7
			G. Ford	19.4	9
			F. Blatz	17.3	2
			D. Batt	32.9	18
$n_1 = 8$	$\Sigma R_1 = 91.5$		$n_1 = 12$	$\Sigma R_2 = 118.5$	

Fourth, apply the procedure for finding the value of U_1. (10.2)

$$U_1 = n_1 n_2 + \frac{n_1(n_1+1)}{2} - \sum R_1 = (8)(12) + \frac{8(8+1)}{2} - 91.5 = 96 + 36 - 91.5 = 40.5$$

Fifth, apply the procedure to find U_2.

$$U_2 = n_1 n_2 - U_1 = (8)(12) - 40.5 = 96 - 40.5 = 55.5 \quad (10.2)$$

Sixth, compare the two U values and select the smaller of the two to use with the z_u. In this example, the smaller value is 40.5.

Seventh, determine the significance of an observed value of U by applying the z_u procedure. (10.3)

$$z_u = \frac{U + \frac{1}{2} - \frac{n_1 n_2}{2}}{\sqrt{\frac{(n_1)(n_2)(n_1 + n_2 + 1)}{12}}} = \frac{40.5 + \frac{1}{2} - \frac{(8)(12)}{2}}{\sqrt{\frac{(8)(12)[8 + 12 + 1]}{12}}} = \frac{-7.5}{\sqrt{\frac{2016}{12}}} = \frac{-7}{12.9615} = -0.54$$

Eight, make a decision. Compare the computed z_u value to the critical value. The computed value is –0.5786, which is between –0 and –1.645. The hypothesis of no difference is accepted.

Minitab Example

1. State the research hypothesis (H_a) and null hypothesis (H_o) in Word.
2. Open Minitab and label two columns as Private and Corporate.
3. Enter the values in the database:

Private	**Corporate**
22.4	21.9
17.8	16.8
26.5	28.0
19.3	19.5
18.2	18.2
21.1	17.9
19.7	35.8
43.5	20.5
	18.7
	19.4
	17.3
	32.9

4. Click on Stat/Nonparametric/Mann-Whitney.

5. Click in the First Sample window; then double-click on Private.
6. Click in the Second Sample window; then double-click on Corporate.
7. Leave the confidence interval at 95.0; this means that the test is at the 0.05 level of confidence.
8. Leave the Alternative as Not Equal.
9. Click on OK.
10. Copy the results from Minitab and paste them below the hypotheses in Word.
11. Summarize the results using the format below:

H_a There was a statistically significant difference in annual income between private and corporate lawyers.
H_o There was no statistically significant difference in annual income between private and corporate lawyers.

Mann-Whitney Test and CI: Private, Corporate

```
             N  Median
Private      8   20.40
Corporate   12   19.45


Point estimate for ETA1-ETA2 is 0.75
95.1 Percent CI for ETA1-ETA2 is (-2.70,4.60)
W = 91.5
Test of ETA1 = ETA2 vs ETA1 not = ETA2 is significant at 0.5892
The test is significant at 0.5890 (adjusted for ties)
```

Since the level of significance was greater than 0.05 (p=0.589), the null hypothesis was accepted and the research hypothesis was not supported. There was no statistically significant difference in annual income between private and corporate lawyers.

Exercises

1. Participants in a Statistics course were divided into two sections. One section did all the problems by hand; the other section did all their computations by computer. At the end of the course after the final exam, the instructor wanted to know if the performance of the two sections on the exam was significantly different. Scores for the sample of the “by hand” section were 50, 48, 47, 45, and 41. Scores for the sample of the “computer” section were 51, 49, 45, 45, and 43. Because the sample was too small to justify using a t-Test, Mann-Whitney was the correct procedure even though the data were interval. Using Mann-Whitney,

analyze the difference between the scores of the two sections. Report the hypotheses, Minitab printout, and conclusion using the format for the general example above.

2. Use the income amounts of the eight private lawyers from the general example above and the income amounts of the first eight corporate lawyers (leave off the 18.7, 19.4, 17.3, and 32.9). Change the background story to the following: Eight private lawyers became dissatisfied with their annual income and switched to corporate law. Analyze whether there was a net gain or loss in their annual income using the appropriate statistical test by answering the four questions and consulting the Decision Model. Report the hypotheses, Minitab printout, and conclusion in the format for the general example above. Assume the data to be normally distributed.

CHAPTER 11

KRUSKAL-WALLIS

Basic Concepts

The Kruskal-Wallis[26] one-way analysis of variance by ranks is a useful test for determining whether three or more columns of values[27] are from different populations when the conditions for an ANOVA are not met. Sample values almost invariably differ and the question is whether the differences among the samples signify genuine population differences or whether they represent merely chance variations such as are to be expected among several random samples from the same population (Siegel 1956).

The observations from all columns are combined and ranked in a single series. The smallest score is replaced by rank 1, the next to smallest by rank 2 and so on. When all observations have been ranked, the sum of the ranks in each column is calculated. The Kruskal-Wallis test determines whether the sums of ranks are so disparate that they are not likely to have come from samples, which were drawn from the same population (Siegel 1956).

The Kruskal-Wallis test is similar to the parametric One Way Analysis of Variance or ANOVA in that three or more groups are tested and the test identifies whether there is significant difference without identifying where the difference lies. The Kruskal Wallis test is an extension of the Mann-Whitney U test.

---

[26] Named after W.H. Kruskal and W. A. Wallis. Presented in a paper published in Volume 47 (1952) of the Journal of the American Statistical Association entitled *Use of Ranks in One-Criterion Variance Analysis.*

[27] Sometimes referred to as k independent samples.

Requirements

The conditions necessary for applying the Kruskal-Wallis test are the following: (1) The data must be ordinal in nature. (2) There are three or more groups. (3) The samples are independently drawn. (4) Each sample must contain at least 6 cases (Levin 1983).[28] (5) The data has been changed from scores to ranks.

Procedure

First, state the null hypothesis.

Second, determine the critical value. The degrees of freedom are k – 1.

Third, rank-order the values and find the sum of ranks for each column. Find the total number of values, *N,* by adding $n_1 + n_2 \ldots n_k$. Place all values on a sheet of paper with the numbers in each column placed from highest to lowest. On the sheet of paper, make provisions for adding a second column beside each of the original columns. Identify the original columns as X_1, X_2 and so on. To be consistent, the new, added columns can be identified as R_1, R_2 and so on.

Locate the lowest value in either of the columns and place a *1* in the R column for that sample. Find the second lowest score and place a *2* in the appropriate R Column. This process continues until all scores have been assigned a rank. The last rank should be the value earlier determined to be *N.*

When a value occurs more than once, assign a rank for each occurrence, add the ranks together and divide by the number of tied values. Place the resulting value beside each of the tied values. The next value will be assigned the rank following the last one used in determining the value for the tied values.

[28] When the sample size is less than 6, the researcher is referred to Siegel (1956, 186-188).

Fourth, sum each of the R columns. Find ΣR_1, ΣR_2 and so on. Check the accuracy of the sums of the R columns. The sum of ΣR_1, ΣR_2 and so on should equal $N(N+1)$ divided by 2. If this is not the case, an error has been made and should be corrected.

Fifth, apply the following procedure. (11.1)

$$X_H^2 = \frac{12}{N(N+1)}\left[\frac{\left(\sum_{R1}\right)^2}{n_1} + \frac{\left(\sum_{R2}\right)^2}{n_2} + \ldots \frac{\left(\sum_{RK}\right)^2}{n_K}\right] - 3(N+1)$$

N = Total observations in all groups combined

n = Number of observations in each group

R = Rank of observations

Fourth, the computed X_H^2 is compared with the tabled X_H^2 .05 column. The degrees of freedom associated with X_H^2 are $k - 1$ and are read down the side of the table. If the computed X_H^2 is larger than the tabled X_H^2 the null hypothesis of no difference is rejected. If the computed X_H^2 is smaller than the table X_H^2, the null hypothesis is accepted.

If the hypothesis is accepted, the samples came from the same population. Rejection of the hypothesis brings a need to identify where the differences lie. Which group or groups are significantly different from another or other groups?

The Mann-Whitney *U* Test is applied two columns at a time to locate differences. Begin by dividing each ΣR by the *n* for that sample. This process is important if the samples differ greatly in size. Arrange the Rs in order from the highest to the lowest. Find the difference between each of the average Rs and order the resulting differences from highest to lowest.

The Mann-Whitney Test should be applied to each of the pairs of comparisons. In order to make the comparisons, return to the original values and ranks. The Mann-Whitney Test continues from the highest comparison to the next highest comparison as long as the computed Z_U is greater than the tabled Z_U of 1.645, which is the point of rejection for the hypothesis of equality of the samples. The first comparison that produces a computed ZU less than the tabled ZU ends the application of the Mann-Whitney Test. All remaining comparisons will produce an acceptance of the hypothesis of equality.

General Example

Ron, Ted, and Sam are hotshot salesmen of new wholesale farm machinery for Righteous Brothers Right Deal, Inc. Almost Righteous, managing partner, is interested in testing the hypothesis that there is no difference in the mean monthly commission of the three hotshot salesmen. Commissions for 8 months were chosen as representative monthly draws for the three. The commissions were recorded in thousand-dollar units.

First, state the null hypothesis. Among the three hotshot salesmen, there is no significant difference in commission.

Second, find the critical value. The degrees of freedom is k – 1 = 2. The table value of X_H^2 at the 0.05 level with 2 degrees of freedom is 2.394.

Third, develop a worksheet in which to pool and rank the commissions. The ranks and numbers from which the ranks were taken were divided into separate columns and totaled (See Table 29).

Table 21 – Kruskal Wallis Ranking Worksheet
Commission Comparison[29]

Month	Ron		Ted		Sam	
	Com.	Rank	Com.	Rank	Com.	Rank
1	2.3	10	2.8	15	2.0	5
2`	2.1	6.5	2.9	16	1.9	4
3	1.5	1	3.1	17.5	2.6	13.5
4	2.3	10	1.8	3	3.9	22
5	4.6	24	1.7	2	4.4	23
6	3.8	20.5	3.4	19	2.2	8
7	3.1	17.5	2.1	6.5	3.8	20.5
8	2.6	13.5	2.3	10	2.5	12
		$\Sigma R_1 = 103$		$\Sigma R_{2=}89$		$\Sigma R_3 = 108$

Fourth, apply the Kruskal Wallis X^2 procedure. (11.1)

$$X_H^2 = \frac{12}{N(N+1)}\left[\frac{\left(\sum_{R1}\right)^2}{n_1}\right]+\left[\frac{\left(\sum_{R2}\right)^2}{n_2}\right]+\left[\frac{\left(\sum_{R3}\right)^2}{n_3}\right]-3(N+1))$$

$$X_H^2 = \frac{12}{24(24+1)}\left[\frac{(103)^2}{8}\right]+\left[\frac{(89)^2}{8}\right]+\left[\frac{(108)^2}{8}\right]-3(24+1)$$

$$= \frac{12}{600}\left[\frac{10{,}609}{8}\right]+\left[\frac{7{,}921}{8}\right]+\left[\frac{11{,}664}{8}\right]-3(24+1)$$

$$X_H^2 = .02[1{,}326.13]+[990.13]+[1458]-75 = .02[3{,}774.26]-75 = 75.49-75 = .49$$

Fifth, make a decision. Compare the computed value of X_H^2 to the table value. The computed value is 0.49; the table value of X_H^2 at the 0.05 level with 2 degrees of freedom is 2.394. Since the tabled value is larger than the computed value, the null hypothesis of no difference is accepted.

[29] Kvanli et al. 1986

Minitab Example

1. State the research and null hypotheses.
2. Enter the data in two columns labeled Commission and Salesman.

   The Worksheet will look like the following:

Commission	Salesman
2.3	Ron
2.1	Ron
1.5	Ron
2.3	Ron
4.6	Ron
3.8	Ron
3.1	Ron
2.6	Ron
2.8	Ted
2.9	Ted
3.1	Ted
1.8	Ted
1.7	Ted
3.4	Ted
2.1	Ted
2.3	Ted
2.0	Sam
1.9	Sam
2.6	Sam
3.9	Sam
4.4	Sam
2.2	Sam
3.8	Sam
2.5	Sam

3. Click on the Stat button in the toolbar. Then click on Non-Parametric and Kruskal-Wallis. Enter Salesman as the Factor and Commission as Response.
4. Click OK.

The outcome appears as follows:

Kruskal-Wallis Test: Commission versus Salesman

Kruskal-Wallis Test on Commissi

Salesman	N	Median	Ave Rank	Z
Ron	8	2.450	12.9	0.18
Ted	8	2.550	11.1	-0.67
Sam	8	2.550	13.5	0.49
Overall	24		12.5	

H = 0.49 DF = 2 P = 0.785
H = 0.49 DF = 2 P = 0.784 (adjusted for ties)

The interpretation of the printout is that since the level of significance was greater than 0.05 (p=0.784), the null hypothesis was accepted and the research hypothesis was not supported. The salesmen's monthly commissions were statistically the same.

Exercises

1. A researcher tested three different brands of breakfast drinks to see how many milliequivalents of potassium per quart each contains. The following samples were obtained randomly: Brand A 4.7, 3.2, 5.1, 5.2, and 5.0; Brand B 5.3, 6.4, 7.3, 6.8, and 7.2; and Brand C 6.3, 8.2, 6.2, 7.1, and 6.6. At the $\alpha = 0.05$. These data are Interval, but the sample is too small to produce a normal distribution; therefore, Kruskal-Wallis is the proper statistical analysis. Is there enough evidence to reject the hypothesis that all brands contain the same amount of potassium? Is there a brand that should be dropped from the product line because of a shortage of potassium? If so, which brand should be dropped? Write the hypotheses and do the procedure with a conclusion.
2. Using the data above, apply the proper procedure chosen by means of

the Decision Model if Brand C were not in the picture at all. Assume a non-normal distribution. Write the hypotheses and do the analysis with a conclusion.

3. Assume a normal distribution using all the data in exercise number 1 above, use the proper parametric statistical test and report the results as you did in 1 and 2 above.

CHAPTER 12

WILCOXON SIGNED-RANKS TEST

Basic Concept

The Wilcoxon Signed-Ranks Test is a nonparametric test for comparing two related samples.[30] The parametric counterpart is the Paired t-Test. The test often has the words *matched pairs* in the title. Each value is matched in one sample with a value in the second sample[31]. The pairs are matched in order to produce a difference value. A consistent pattern where most difference scores are of the same sign will indicate significance. A mixture of equal positive and negative differences indicates a lack of consistency in the differences between the samples. Tied scores, that is a difference score of zero, are dropped from the analysis and *n* is reduced by 1 (Hinton 1996).

In addition to comparing the sign of the differences, the Wilcoxon test also compares the size of the differences. The test ranks the size of the differences, that is their absolute values, by ignoring the sign of the differences and treating them all as positive for ranking purposes (Hinton 1996).

If the computed z_w is outside –1.645 to +1.645, the hypothesis of equality is rejected. If the computed z_w is within the critical value of –1.645 to +1.645, the hypothesis of equality is accepted.

Procedure

First, state the null hypothesis.

---

[30] Spatz (1997) describes three types of correlated-samples – natural pairs, matched pairs and repeated measures. For a more complete discussion see pages 190-191.

[31] The Mann-Whitney test assumes independent samples. The Wilcoxon Signed-Rank Test for Paired Samples includes such samples as husband-wife, brother-sister, and before-after combinations.

Second, determine the critical value. If the computed z is larger than 1.645 or smaller than –1.645 the hypothesis of no difference is rejected.

Third, develop a table in which to subtract the initial measurement from a later measurement to obtain a difference value for each respondent. When a respondent has a zero difference value, remove the observation from the analysis, and reduce the size of *n* by 1.

Fourth, rank the resulting difference values from lowest to highest, ignoring the positive or negative signs. Assign a rank for each occurrence of tied or repeated values then add the ranks together and divide by the number of tied scores. Place the resulting rank besides each tied score.

Fifth, sum the ranks of the positive differences, ΣR_+, and sum the ranks of the negative differences, ΣR_-. The smaller of the positive and negative sums of ranks is the calculated value of *t* and is used during the application of the z_w procedure.

Sixth, apply the Wilcoxon z procedure. (12.1)

$$Z_W = \frac{t - \frac{N(N+1)}{4}}{\sqrt{\frac{N(N+1)(2N+1)}{24}}}$$

t = value of the smallest of the two signed values.

N = number of observations in which a difference between the first and second observations is noted.

If the computed z_w is larger than 1.645, the hypothesis of equality is rejected. If the computed z_w is smaller than –1.645, the hypothesis of equality is rejected and if the computed z_w is between +1.645 and –1.645 the hypothesis of equality is accepted.

General Example

Dr. Feel Better, a noted psychologist, offered a seminar to increase personal esteem. A before-and-after test was administered to measure self-esteem before and after the seminar. The before-and-after test values of nine representative participants were analyzed to determine if there were significant gains in self-esteem. The before-and-after values were 70-74; 72-88; 75-71; 61-62; 82-89; 55-58; 43-41; 51-63; and 84-80.

First, state the null hypothesis. Among seminar participants, there was no significant difference in the before and after measurement of self-esteem.

Second, determine critical value. If the computed z is larger than 1.645 or smaller than –1.645 the hypothesis of no difference is rejected.

Third, develop a table in which to subtract the initial measurement from a later measurement to obtain a difference value for each respondent. When a respondent has a zero difference value, remove the observation from the analysis, and reduce the size of *n* by 1 (See Table 31).

Fourth, rank the resulting difference values from lowest to highest, ignoring the positive or negative signs. Assign a rank for each occurrence of tied or repeated values then add the ranks together and divide by the number of tied scores. Place the resulting rank besides each tied score.

Fifth, sum the ranks of the positive differences, ΣR_+, and sum the ranks of the negative differences, ΣR_-. The smaller of the positive and negative sums of ranks is the calculated value of *t* and is used during the application of the z_w formula.

Table 22 -- Wilcoxon Self-Esteem Analysis Worksheet

X_1	X_2	Diff.	Ranks	R_+	R_-
70	74	+4	5	+5	
72	88	+16	9	+9	
75	71	-4	5		-5
61	62	+1	1	+1	
82	89	+7	7	+7	
55	58	+3	3	+3	
43	41	-2	2		-2
51	63	+12	8	+8	
84	80	-4	5		-5
				$\Sigma R_+ = 33$	$\Sigma R_- = 12$

Sixth, apply the Wilcoxon z procedure (12.1)

$$Z_W = \frac{t - \frac{N(N+1)}{4}}{\sqrt{\frac{N(N+1)(2N+1)}{24}}}$$

$$Z_W = \frac{12 - \frac{18(18+1)}{4}}{\sqrt{\frac{18(18+1)(36+1)}{24}}} = \frac{12 - \frac{342}{4}}{\sqrt{\frac{(342)(37)}{24}}} = \frac{12 - 85.5}{\sqrt{\frac{12{,}654}{24}}} = \frac{-73.5}{\sqrt{527.25}} = \frac{-73.5}{22.96} = 3.20$$

Seventh, make a decision. Compare the calculated z value to the table value. The computed z_w is larger than 1.645, the hypothesis of equality is rejected. If the computed z_w had been between a –1.645 and a +1.645 the hypothesis of equality would have been accepted (Field unpublished).

MINITAB Example

The same example used above will demonstrate the ease and quickness of the MINITAB program. Dr. Feel Better, a noted psychologist, offered a seminar to increase personal esteem. A before-and-after test was administered to measure self-esteem before and after the seminar. The before-and-after test values of nine representative participants were analyzed to determine if there were significant gains in self-esteem. The before-and-after scores were: 70-74; 72-88; 75-71; 61-62; 82-89; 55-58; 43-41; 51-63; and 84-80.

1. State the research and null hypotheses.

   H_a Among seminar participants, there was a statistically significant difference in the before and after measurements of self-esteem.

   H_o Among seminar participants, there was no statistically significant difference in the before and after measurements of self-esteem.

2. Establish a database in Minitab by opening the program and labeling two columns (Before and After).
3. In the toolbar click Calc and Calculator. Type ‘c3’ in the “Store Results in Variable” box.
4. In the “Expression” Box enter ‘C1-C2’ (this means C1 minus C2). Click OK. Now column C3 has the difference between the Before and the After scores. When the data are entered, it is time to analyze.
5. Click on Stat, Nonparametric, and 1-Sample Wilcoxon. Enter C3 in the Variables window. Be sure to check the box marked Test Median. Click OK.

The resulting printout looks like the following:

Wilcoxon Signed Rank Test: C3

```
Test of median = 0.000000 versus median not = 0.000000

        N for   Wilcoxon              Estimated
     N   Test  Statistic        P        Median
C3   9      9       12.0    0.236       -3.500
```

Since the level of significance was greater than 0.05 (p=0.236), the null hypothesis was accepted and the research hypothesis was not supported. Dr. Feel Better's self-esteem seminar did not increase self-esteem.

Exercises

1. In a large department store chain, the owner wishes to see whether the number of shoplifting incidents per day will change if the number of uniformed security officers is doubled. From a random sample of six stores seven days before security was increased there was a total of 63, 12, 18, 36, 30, 48, and 72 incidents of shoplifting. In seven days after the security was increased there was a total of 30, 18, 24, 18, 6, 36, and 24 incidents of shoplifting in the sample of stores in the chain. Even though the data are interval, the sample size is too small to use the Paired *t*-Test. Is there enough evidence to support the claim, at α = 0.05, that there is a difference in the number of shoplifting incidents before and after the increase in security? The cost of the doubled security will be between $150,000 and $200,000 for the whole chain. Write the hypotheses and analyze the data as in the example above; report the results accordingly.
2. Assume the data above in exercise 1 to be a normal distribution and

run the correct parametric test. Report the results in the normal format with hypotheses, data, and conclusion.

CHAPTER 13

SPEARMAN RANK CORRELATION

Basic Concept

The Spearman Rank Correlation Coefficient[32], symbolized by r_s, is an adaptation of the Person *r* Correlation Coefficient for use with variables measured in the form of ranks. When the Pearson Product Moment of Correlation requirements, especially the requirement for normal distribution are unmet, the Spearman Rank Correlation Coefficient can be used. Spearman Ranks Correlation determines if two measures taken on a single group produces a significant relationship between the two measures. Both Spearman r_s and Pearson *r* have values ranging from –1 to + 1.

If the ranks agree perfectly with each other, differences in the ranks would be zero and rho would be 1.0, indicating a perfect positive association between two variables. If the two series of rankings were in exactly reverse order, the value of rho would be –1, indicating a perfect negative association.

Procedure

First, state the null hypothesis.

Second, determine the critical value.

Third, develop a worksheet to calculate ΣD^2. The two sets of measurements, X_1 and X_2, are placed on a sheet of paper in such a way that columns, R_1 and R_2, can be added for ranking each set of measurements. Rank

[32] Charles Edward Spearman, a British psychologist, developed a measure of association called the rank-difference correlation coefficient. (Tai 1978)

the values from lowest to highest with X_1 and X_2 in columns R_1 and R_2. The rank numbers will be from 1 to a number equal to the number of cases.

Fourth, subtract the ranks in R_1 from the ranks in R_2 and place the amount in a new column D. If R_1 is larger than R_2, then D will be a + value. If R_1 is smaller than R_2, then D will be a - value.

Fifth, square the values in column D in a new column labeled D^2.

Sixth, sum the D^2 values to obtain ΣD^2.

Seventh, calculate r_s with the following procedure:

$$r_s = 1 - \frac{6\sum D^2}{N\left(N^2 - 1\right)} \qquad (13.1)$$

N = number of respondents

D^2 = differences square

Eighth, make a decision. Test the hypothesis that $r_s = 0$. The test is made with z_{rs} as follows:

$$z_{rs} = r_s\sqrt{N - 1} \qquad (13.2)$$

r_s = computed value based upon the square of differences

N = number of respondents or groups

df = n – 2

If the computed z_{rs} lies between 0 and the tabled value of –1.645, the hypothesis, $z_{rs} = 0$, is accepted. A table value between 0 and –1.645 indicates that neither the –1 nor the +1 condition dominates to the extent necessary for the hypothesis to be rejected. If the calculated z_{rs} is outside the tabled value of 0 to –1.645, the null hypothesis is rejected and a significant correlation is assumed.

General Example

Two judges ranked ten participants in the Bugtussel National Opossum Baking Contest. The officials of the contest evaluated the performance of the two judges. They were particularly interested in knowing if the two judges were significantly different in their ranking of the participants.

First, state the hypothesis. Between the two judges, there is no significant difference in their ranking of participants.

Second, determine the critical value. The tabled value of rho (see Appendix 8, Spearman r_s) is 0.648 at the 0.05 level for a two-tailed test with 10 pairs. If the calculated value is equal to or greater than 0.648, the null hypothesis of no difference is rejected and a significant correlation is assumed.

Third, develop a worksheet to calculate ΣD^2.

Table 23 -- Spearman Rank Correlation Worksheet
Baking Contest

Cooks	Judge 1 X_1	R_1	Judge 2 X_2	R_2	D $(R_1 - R_2)$	D^2
A	2	9	6	5	+4	16
B	6	5	2	9	-4	16
C	10	1	9	2	-1	1
D	4	7	7	4	+3	9
E	8	3	10	1	+2	4
F	5	6	3	8	-2	4
G	1	10	4	7	+3	9
H	7	4	5	6	-2	4
I	3	8	1	10	-2	4
j	9	2	8	3	-1	1
			ΣD^2=68			

Fourth, apply the following procedure: (13.1)

$$r_s = 1 - \frac{6\sum D^2}{N(N^2-1)} = 1 - \frac{6(68)}{10(100-1)} = 1 - \frac{408}{990} = 1 - .4121212 = 0.5879$$

Fifth, compare the calculated r_s of 0.5879 with the Table of Critical Values for Spearman (Appendix 8).

Since the calculated r_s of 0.5879 is not equal to or larger than the table value at 0.05 for a two-tailed test with 10 pairs, the null hypothesis is accepted. The correlation is not significant.

MINITAB Example

1. State the research and null hypotheses:

   H_a There was a statistically significant correlation between the rankings of the two judges.

   H_o There was not a statistically significant correlation between the rankings of the two judges.

2. Open a database in Minitab and enter the data in two columns labeled Judge 1 and Judge 2.

3. Manipulate the data in order to run Spearman in Minitab.

   a) Click on Data/Stack/Columns.

   b) In "Stack the following columns" box enter C1 and C2.

   c) Click on "Column of current worksheet" enter C3

   d) In "Store Subscripts in" enter C4

   e) Label C3 "Score" and C4 "

   f) Now you have all the data ready for ranking. Click on Data in the toolbar, and then click on Rank. Enter C3 in the box for Rank Data In; enter C5 for Store Ranks In. Highlight and copy rows 11-20 and paste them in C6. Label C5 as Rank Judge 1, and label C6 as

Rank Judge 2.

4. Fourth, click on Stat, Basic Statistics, and Correlation. Under 'Variables' enter Rank Judge 1 and Rank Judge 2. Be sure the Display p-values box is checked; then click OK.

The resulting printout is as follows:

Correlations: Rank Judge 1, Rank Judge 2

Pearson correlation of Rank Judge 1 and Rank Judge 2 = 0.588
P-Value = 0.074

Minitab gives the Pearson correlation, but since it is based on ranked data, the result is actually the Spearman's Rank Correlation. Change "Pearson" in the printout to "Spearman." Minitab does give the P-Value (in this case it was 0.074); so the analysis is complete. Since the level of significance was greater than 0.05 (p-0.074), the null hypothesis was accepted and the research hypothesis was not supported. There was not a statistically significant correlation in the rankings of the two judges.

Exercises

1. Two statistics instructors were asked to rate eight statistics textbooks on an ascending scale from 0 to 20 points. Points were assigned for reading level, use of illustrations, and use of color. Instructor 1 rated the textbooks 4, 10, 18, 20, 12, 2, 5, and 9. Instructor 2 rated the textbooks 4, 6, 20, 14, 16, 8, 11, and 7. At $\alpha = 0.05$, test the hypothesis that there was a statistically significant linear correlation between the two instructor ratings. Use the proper format with hypotheses, data, and conclusion.
2. Assume that the data in exercise 1 above are ratings of ten students before and after the stats class. Determine the proper statistical procedure to use with the Decision Model and run the test using the proper format of hypotheses, data, and conclusion.

CHAPTER 14

FRIEDMAN TWO-WAY ANALYSIS OF VARIANCE TEST

Basic Concept

The Friedman Two-Way Analysis Of Variance By Ranks Test, symbol X_F^2, is useful for testing the null hypothesis that three or more columns were drawn from the same population. The data are three or more measurements from matched columns that are at least ordinal level measurements. The Friedman Two-Way Analysis Of Variance By Ranks Test is a nonparametric approach to test differences in a single group that was measured under three or more conditions. The Friedman provides analysis and comparison of three or more nonparametric measurement.

Requirements

In order to apply the Friedman two-way analysis of variance, the following conditions must be satisfied: (1) Ordinal data capable of being ranked are required. The data need not be normally distributed with equal variances between repeated measures. (2) One group or a single sample is required. Members of two or more groups may be measured as one group if matched on relevant variables. (3) The number of respondents must not be too small. Minimal requirements for *N* depends on the number of conditions or *k* to which respondents are exposed.[33] (4) The measurement is applied to the same

[33] See Siegel (1956, 168-169) for detailed instructions for a small number of respondents. Siegel (1956, 280-281) gives exact probabilities associated with values of $k = 3$, $N = 2$ to 9, and for $k = 4$, $N = 2$ to 4. For N and/or k larger than those shown in these tables, the associated probability may be determined by reference to the chi-square distribution with $df = k - 1$.

sample of respondents three or more times. Three or more measurement of the same sample meets the required condition of matched or related data. The Kruskal-Wallis Test is similar but does not require matched samples. (5) The raw values must be changed to ranked data.

Procedure

First, state the null hypothesis.

Second, determine the critical value. The degrees of freedom are read down the side of the table. If the computed X_F^2 is larger than the tabled X^2 the hypothesis of equality is rejected. If the computed X_F^2 is smaller than the tabled X_F^2 the hypothesis of equality is accepted.

Third, develop a worksheet in which the scores are placed in a two-way table having k columns representing the conditions and N rows representing the subjects or groups. Rank the scores in each row from 1 to k in new columns identified as R_1, R_2, R_3 and so on until the total number of measurements or k samples have been identified. If three measurements are recorded the rankings within a row will be 1, 2 and 3 and ranked values will be from the smallest to largest. Because of the repeated measurements, the values are ranked within each row rather than across all the scores. Sum the ranks in each column to obtain ΣR. Square each ΣR to obtain $(\Sigma R)^2$. Sum the $(\Sigma R)^2$ to obtain the $\Sigma(\Sigma R_k)^2$.

Fourth, compute the value of X_F^2, using the values obtained in step 3 and the following procedure.

$$X_F^2 = \frac{12}{N_k(k+1)}\left[\sum\left(\sum R_k\right)^2\right] - 3N(k+1) \qquad (14.1)$$

N = the number of rows

k = the number of columns

df = k – 1

All components of the formula are constant except $\Sigma(\Sigma R_k)^2$.

Fifth, make a decision. Compare the computed X_F^2 to the 0.05 column of the tabled X^2. The degrees of freedom are read down the side of the table[34]. If the computed X_F^2 is larger than the tabled X^2 the hypothesis of equality is rejected. If the computed X_F^2 is smaller than the tabled X_F^2 the hypothesis of equality is accepted. If the hypothesis is accepted, there is no significant difference among the groups and it is assumed that the groups came from the same population. If the hypothesis is rejected, there may be need to determine where the significant difference lies.

General Example[35]

The Rotweiler Raunchy Rat Ranch studied three patterns of reinforcement upon discrimination learning in three matched samples (k = 3) of 18 groups of raunchy rats. Three patterns of reinforcement were administered. Pattern A was training with 100% reinforcement; pattern B was training with partial reinforcement, and pattern C was training with an altered form of partial

---

[34] See Siegel (1956, 168-169) for detailed instructions for a small number of respondents. Siegel (1956, 280-281) gives exact probabilities associated with values of $k = 3$, $N = 2$ to 9, and for $k = 4$, $N = 2$ to 4. For N and/or k larger than those shown in these tables, the associated probability may be determined by reference to the chi-square distribution with $df = k - 1$.

[3]Numbers were borrowed from Seigel (1956).

reinforcement. Learning was measured by the speed at which the rats learned an opposing habit. The null hypothesis stated no difference in the rating of the number of errors in learning among the three types of reinforcement. The critical value with α = 0.05, N = 18, k = 3, and df = k – 1 was 5.99.

First, state the hypothesis. Among the three types of reinforcement, there was no significant difference in the ratings of the number of errors.

Second, determine the critical value. The X^2 table value at the 0.05 level with 2 degrees of freedom was a value of 8.6.

Third, prepare a worksheet in which the number of errors was placed in a table with 3 columns and 18 rows. Each row of 3 rats was ranked within the row.

Table 24 -- Friedman Two-Way Analysis of Variance Worksheet

	Type of reinforcement		
Group	A	B	C
1.	1	3	2
2.	2	3	1
3.	1	3	2
4.	1	2	3
5.	3	1	2
6.	2	3	1
7.	3	2	1
8.	1	3	2
9.	3	1	2
10.	3	1	2
11.	2	3	1
12.	2	3	1
13.	3	2	1
14.	2	3	1
15.	2.5*	2.5*	1
16.	3	2	1
17.	3	2	1
18.	2	3	1
R_k=	39.5	42.5	26.0

* Animals obtained equal scores

The sum of the ranks for pattern A was calculated at 39.5; for pattern B at 42.5, and for pattern C at 26.0

Fourth, calculate the value of X^2_F.

$$\begin{aligned} X_F^2 &= \frac{12}{Nk(k+1)}\left[\sum\left(\sum R_k\right)^2\right] - 3N(k+1) \\ &= \frac{12}{[(18)(3)](3+1)}\left[(39.5)^2 + (42.5)^2 + (26.0)^2\right] - 3(18)(3+1) \\ &= \frac{12}{216}[1560.25 + 1806.25 + 676] - 216 = 0.0555555[4{,}042.5] - 216 \\ &= 224.5831 - 216 = 8.6 \end{aligned} \tag{14.1}$$

Fifth, make a decision. Compare the X^2 table value (see Appendix 5, Chi-Square Distribution) at the 0.05 level with 2 degrees of freedom of 5.99 with the computed value of 8.6. The null hypothesis of no difference was rejected. There was a statistically significant difference among the three patterns of reinforcement.

MINITAB Example

1. State the null hypothesis: Among the three patterns of reinforcement, there was no statistically significant difference in the ratings of the number of errors using the eighteen blocks of rats.

2. Open Minitab and enter the data in three columns: Group, Errors, and Reinforcement. In the general example each Group must be entered three times (1, 1, 1); then the data for group one are entered under Errors (1, 3, 2); then the Reinforcement is entered (A, B, C). Fifty-four rows of Minitab will be used when this is completed.

3. Analyze the data by clicking on 'Stat'/'Nonparametrics'/'Friedman.'

4. In the Friedman window enter Errors under Response, Reinforcement

under Treatment, and Group under Blocks. Click OK.

The results will be as follows:

Friedman Test: Errors versus Treatment blocked by Group

```
S = 8.58  DF = 2  P = 0.014
S = 8.70  DF = 2  P = 0.013 (adjusted for ties)

                              Sum of
Treatment    N  Est Median   Ranks
A           18      2.0000    39.5
B           18      2.6667    42.5
C           18      1.3333    26.0

Grand median = 2.0000
```

Since the level of significance was less than 0.05 (p=0.013), the null hypothesis was rejected and the research hypothesis was supported. There was a statistically significant difference in the ratings of the number of errors among the three patterns of reinforcement using the eighteen blocks of rats. Reinforcement C seemed to have fewer errors than either A or B.

Exercise

In a study of the attractiveness of new car warranties, a newspaper writer sent the warranties of three competing brands of cars (Kia, Hyundai, and Toyota) to 10 different editors of car magazines. Each editor was asked to rank the three warranties. The warranties were assigned in a random manner for each editor. The results were:

Editor	Kia	Hyundai	Toyota
1	3	2	1
2	2	1	3
3	3	1	2

4	2	1	3
5	3	2	1
6	2	1	3
7	1	2	3
8	2	1	3
9	3	1	2
10	1	2	3

Use the Friedman test to determine if there was a difference in the perceived quality of the three warranties. The data setup in Minitab will use three columns and 30 rows.

CHAPTER 15

CHI-SQUARE GOODNESS OF FIT

Basic Concept

The Chi-Square Goodness of Fit Test[36], also called one-sample chi-square test, measures a set of observed frequencies to see if there is a good fit for an expected frequency pattern. The test also measures if the sample results are consistent with the results obtained when a random sample is selected from a population with a known distribution. The number of categories may be two or more.

If the agreement between the observed and expected frequencies is close, the differences will be small and consequently X^2 will be small. If the divergence is large, the value of the computed X^2 will be large. The larger X^2 becomes the more likely the observed frequencies did not come from the population on which the null hypothesis is based. When the computed value is larger than the table value, the null hypothesis is rejected.

Requirements

Several requirements are set forth for the use of the X^2 Goodness of Fit Test including, (1) two or more independent samples are needed for comparison, a comparison of related data is inappropriate; (2) nominal data, with frequencies only, are required; (3) the data observations must be independent and each data value can be counted in one and only one category; (4) the sample should be

[36] Karl Pearson (1857-1936) first used the chi-square distribution as a goodness-of-fit test for data.

drawn at random from the population, and (5) the expected frequencies should not be too small.[37]

Procedure

First, state the null hypothesis.

Second, determine the critical value. Find the degrees of freedom – degrees of freedom = (k – 1)(r – 1). Compare the calculated X^2 value with the table X^2 value.

Third, develop a contingency table by arranging the *observed and expected frequencies* in a contingency table. Subtract the *expected frequencies* from the *obtained frequencies,* (O – E). Square the *differences* between the expected and obtained frequencies, $(O - E)^2$.

Fourth, divide the squared difference by the observed frequency minus the expected frequency, $\frac{(O-E)^2}{E}$.

Fifth, sum the quotients, $X^2 = \sum \frac{(O-E)^2}{E}$ (15.1)

Sixth, find the degrees of freedom – $df = (k - 1)(r - 1)$.

Compare the calculated Chi-Square with the tabled Chi-Square. If the obtained Chi-Square is larger than the tabled Chi-Square the null hypothesis is rejected.

[37] "For a 2 x 2 problem, no expected frequency should be smaller than 5. Yates's corrected formula should be used for a 2 x 2 problem in which an expected cell frequency is smaller than 10" (Levin 1983, 186).

General Example

A sociologist wishes to determine if attitudes toward legalized abortion are changing. A survey taken five years previously showed 60% opposed to legalized abortion, 25% indifferent, and 15% in favor. The sociologist takes a new sample of 200 respondents and finds that 105 opposed legalized abortion, 40 were indifferent and 55 favored legalized abortion.

First, state the null hypothesis. Between the measurement of 5 years ago and the current measurement, there is no significant different in the percentages of the three responses.

Second, determine the critical value. The critical value for X^2 at the 0.05 level of confidence and degrees of freedom = $(k - 1)(r - 1)$ = (3-1)(2 – 1) = (2)(1) = 2 degrees of freedom is 5.99. (See Appendix 5)

Third, prepare a contingency table with the observed frequencies and the expected frequencies. Based on percentages of the previous survey: E1 = (200) (.60) = 120; E2 = (200) (.25) = 50; E_3 = (200) (.15) = 30

Table 37 -- Chi-Square Goodness of Fit Analysis
Abortion Attitudes

	Opposed	Indifferent	For	Sample Size
O - Observed	105	40	55	200
E - Expected	120	50	30	200
	225	90	85	400

Fourth, calculate X^2. (15.1)

$$X^2 = \sum \frac{(O-E)^2}{E} = \frac{(105-120)^2}{120} + \frac{(40-50)^2}{50} + \frac{(55-30)^2}{30}$$
$$= \frac{(-15)^2}{120} + \frac{(-10)^2}{50} + \frac{(25)^2}{30} = \frac{225}{120} + \frac{100}{50} + \frac{625}{30} = 1.87 + 2 + 20.83 = 24.70$$

Fourth, compare the calculated X^2 with the table X^2. The critical value for X^2 at the 0.05 level of confidence and degrees of freedom = $(k-1)(r-1) = (3-1)(2-1) = (2)(1) = 2$ degrees of freedom is 5.99. Since the computed value of 24.70 is larger than the table value 5.99, the hypothesis of no difference is rejected. Attitudes toward legalized abortion seem to have changed.

MINITAB Example

1. State the null hypothesis: Between the measurement of 5 years ago and the current measurement, there was no statistically significant difference in the percentages of the three responses.

2. Open Minitab. Label the first column 'Position,' the second column 'Observed,' and the third column 'Ratios.'

3. Enter the data under 'Observed' from the sample. Enter the data under 'Ratios' as decimals (60% is .60, etc.). Enter the data under 'Position" as 'Opposed,' 'Indifferent,' and 'For.'

4. Click on 'Stat,' 'Tables,' and 'Chi-Square Goodness of Fit.' Under 'Observed counts' enter 'C2 Observed.' Under 'Category names' enter 'C1 Position.' Under 'Test' click Specific Proportions and enter 'C3 Rations.' Click OK. The printout is below:

Chi-Square Goodness-of-Fit Test for Observed Counts in Variable: Observed

```
Using category names in Position

                               Test              Contribution
Category       Observed  Proportion  Expected     to Chi-Sq
Opposed             105        0.60       120        1.8750
Indifferent          40        0.25        50        2.0000
In Favor             55        0.15        30       20.8333

  N  DF   Chi-Sq  P-Value
200   2  24.7083    0.000
```

Since the level of significance was less than 0.05 (p=0.000), the null hypothesis was rejected and the research hypothesis was supported. There has been a change in the percentages of responses since the last time data were collected.

Exercise

A market analyst wishes to see if consumers' preferences among five flavors of a fruit soda had changed. Previously consumers preferred the flavors in the following ratios: 15% cherry, 15% strawberry, 25% orange, 20% lime, and 25% grape. A current sample of 160 people provided the following data: cherry – 40, strawberry – 30, orange – 36, lime – 19, and grape – 35. Are the differences significant or are the differences due to chance? Null hypothesis: Consumers show no change in preference for flavors of the fruit soda.

CHAPTER 16

CHI-SQUARE TEST OF INDEPENDENCE

Basic Concept

The Chi-Square Test of Independence, also called the Chi-Square Test of Association, tests whether two variables are independent or associated. The degree to which the observed frequencies differ from the expected frequencies is measured by the X^2. Each variable has two or more mutually exclusive categories. The test rule is to reject the null hypothesis if X^2 is large and accept the hypothesis if X^2 is small. The critical value is determined by the choice of α and the degrees of freedom when consulting a chi-square distribution. A cause-and-effect relationship between two variables is tempting when a chi-square test of independence results in a rejection of the null hypothesis. A rejection of the null hypothesis indicates only that there may be some relationship between the variables.

Requirements

Several requirements are set forth for the use of the X^2 Goodness of Fit Test. (1) Two or more independent samples are compared. A comparison of related data is inappropriate. (2) Nominal data, with frequencies only, are required. (3) The data observations must be independent and each data value can be counted in one and only one category. (4) The sample are randomly drawn from the population. (5) The expected frequencies should not be too small.[38]

---

[38] "For a 2 x 2 problem, no expected frequency should be smaller than 5. Yates's corrected formula should be used for a 2 x 2 problem in which an expected cell frequency is smaller than 10" (Levin 1983, 186).

Procedure

First, state the null hypothesis.

Second, determine the critical value. The degrees of freedom = (r – 1) (k – 1). Compare the calculated X^2 to the table X^2. If the calculated value is greater than the tabled value, the null hypothesis is rejected indicating support for a favorable comparison.

r = rows

k = columns

Third, develop a contingency table with 6 columns and as many rows as observations. The columns should be labeled O for observed, E for expected, observed minus expected (O – E), the square of observed minus expected (O – E)^2 , the squared observed minus expected divided by expected. The observed frequencies are given. The expected frequencies may be calculated.

Fourth, apply the X^2 Test for Independence procedure.

$$X^2 = \frac{\sum (O - E)^2}{E} \tag{16.1}$$

Fifth, make a decision. Compare the calculated X^2 to the table X^2. If the calculated value is greater than the tabled value, the null hypothesis is rejected indicating support for a favorable comparison.

General Example

Hook, Crook, and Bummer are candidates for election to the office of County Treasurer. A poll was taken in the three townships in the county to determine the popularity of the candidates and whether there was a difference between townships. In other words, will each candidate get about the same percentage of support in each township?

First, state the null hypothesis. Among the three candidates for county treasurer, there is no significant difference in support in the three townships.

Second, determine the critical value. The value of $\alpha = 0.05$, and the degrees of freedom = (r – 1)(k – 1) = (3 – 1)(3-1) = 4. The X^2 critical value for $\alpha = 0.05$ and df = 4 is 9.49.

Third, develop a worksheet in which the expected frequencies, the row totals, column totals, and grand total are calculated.

Table 26 – Test for Independence Worksheet
County Treasurer Polling Results

Candidates	Elm City	Oakville	Cedar Town	Row Total
Hook	50	40	35	125
Crook	30	45	25	100
Bummer	20	45	20	85
Column Total	100	130	80	310

The expected frequencies for each cell in the contingency table are calculated by the following formula:

$$E = \frac{(\text{row total})\ (\text{column total})}{\text{grand total}} \qquad (16.2)$$

For example, Bummer's expectation in Oakville Township:

$$E = \frac{(85)\ (130)}{310} = 35.7$$

Fourth, develop a contingency table of observed and expected values.

Table 27 – Test of Independence Contingency Table

Candidates	Elm City		Oakville		Cedartown		RowTotal
	Obs	Exp	Obs	Exp	Obs	Exp	
Hook	50	40.3	40	52.4	35	32.3	125
Crook	30	32.3	45	41.9	25	25.8	100
Bummer	20	27.4	45	35.7	20	21.9	85
Column Total	100		130		80		310

Fifth, calculate X^2 Test of Independence.

$$X^2 = \sum \frac{(O-E)^2}{E} \qquad (16.1)$$

The following table implements the procedure.

Table 28 -- X^2 Implementation Worksheet
County Treasurer Polling Results

Row/Column	Observed	Expected	O-E	$(O-E)^2$	$(O-E)^2/E$
1-1	50	40.3	9.7	94.09	2.33
1-2	40	52.4	-12.4	153.76	2.93
1-3	35	32.3	2.7	7.29	0.23
2-1	30	32.3	-2.3	5.29	0.16
2-2	45	41.0	3.1	9.61	0.32
2-3	25	25.8	-0.8	0.64	0.02
3-1	20	27.4	-7.4	54.75	2.00
3-2	45	35.7	9.3	86.44	2.42
3-3	20	21.9	-1.9	3.61	0.16
	310	310.0	-0-		10.48

Sixth, make a decision. Compare the computed value of X^2, 10.48, with the table value of X^2 value 9.49. Since the computed value is larger than the table value, the null hypothesis is rejected. The population percentage favoring each candidate for County Treasurer does not seem to be the same from town to town.

MINITAB Example

The same example used above will demonstrate the Minitab program. Hook, Crook, and Bummer are all candidates for election to the office of County Treasurer. A poll was taken in the three townships in the county to determine the popularity of the candidates and whether or not there was a difference between townships. In other words, will each candidate get about the same percentage of support in each township?

First, state the null hypothesis: among the three candidates for county treasurer, there is no significant difference in support in the three townships.

Second, open Minitab and enter the data in three columns labeled Elm City, Oakville, and Cedar Town. Enter the data for Hook, Crook, and Bummer in rows 1-3.

Third analyze the data by clicking Stats on the Toolbar. Then click Tables and Chi-Square. Enter the columns containing the data and click OK.

The resulting printout will appear as follows:

Chi-Square Test: Elm City, Oakville, Cedar Town

```
Expected counts are printed below observed counts

      Elm City Oakville Cedar Town    Total
    1       50       40         35      125
         40.32    52.42      32.26

    2       30       45         25      100
         32.26    41.94      25.81

    3       20       45         20       85
         27.42    35.65      21.94

Total      100      130         80      310

Chi-Sq =  2.323 +  2.942 +  0.233 +
          0.158 +  0.224 +  0.025 +
          2.008 +  2.455 +  0.171 = 10.539
DF = 4, P-Value = 0.032
```

Since the P-Value is less than 0.05, there is a statistically significant difference in the attitudes towards our three candidates in the three cities. The null hypothesis is rejected.

Exercise

A sociologist wishes to see if the number of years of college is related to place of residence. A sample of 70 people was selected and classified. In urban locations, 15 had no college, 12 had a 4-year degree, 8 had advanced degrees. In Suburban locations, 8 had no college, 15 had 4 year degrees, 9 had advanced degrees. In rural locations, 6 had no college, 8 had 4-years degrees and 7 had advanced degrees. Are the percentages among urban, suburban, and rural residences different?

APPENDIX 1

t DISTRIBUTION

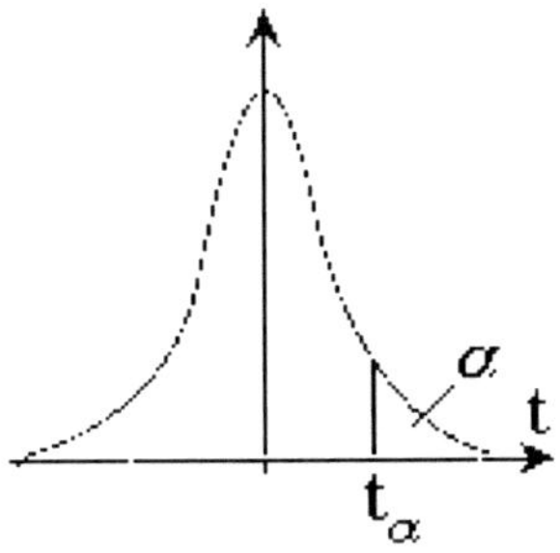

df	α = 0.1	0.05	0.025	0.01	0.005	0.001	0.0005
∞	t_α=1.282	1.645	1.960	2.326	2.576	3.091	3.291
1	3.078	6.314	12.706	31.821	63.656	318.289	636.578
2	1.886	2.920	4.303	6.965	9.925	22.328	31.600
3	1.638	2.353	3.182	4.541	5.841	10.214	12.924
4	1.533	2.132	2.776	3.747	4.604	7.173	8.610
5	1.476	2.015	2.571	3.365	4.032	5.894	6.869
6	1.440	1.943	2.447	3.143	3.707	5.208	5.959
7	1.415	1.895	2.365	2.998	3.499	4.785	5.408
8	1.397	1.860	2.306	2.896	3.355	4.501	5.041
9	1.383	1.833	2.262	2.821	3.250	4.297	4.781
10	1.372	1.812	2.228	2.764	3.169	4.144	4.587
11	1.363	1.796	2.201	2.718	3.106	4.025	4.437
12	1.356	1.782	2.179	2.681	3.055	3.930	4.318
13	1.350	1.771	2.160	2.650	3.012	3.852	4.221
14	1.345	1.761	2.145	2.624	2.977	3.787	4.140
15	1.341	1.753	2.131	2.602	2.947	3.733	4.073
16	1.337	1.746	2.120	2.583	2.921	3.686	4.015
17	1.333	1.740	2.110	2.567	2.898	3.646	3.965
18	1.330	1.734	2.101	2.552	2.878	3.610	3.922
19	1.328	1.729	2.093	2.539	2.861	3.579	3.883
20	1.325	1.725	2.086	2.528	2.845	3.552	3.850
21	1.323	1.721	2.080	2.518	2.831	3.527	3.819
22	1.321	1.717	2.074	2.508	2.819	3.505	3.792
23	1.319	1.714	2.069	2.500	2.807	3.485	3.768
24	1.318	1.711	2.064	2.492	2.797	3.467	3.745
25	1.316	1.708	2.060	2.485	2.787	3.450	3.725
26	1.315	1.706	2.056	2.479	2.779	3.435	3.707
27	1.314	1.703	2.052	2.473	2.771	3.421	3.689
28	1.313	1.701	2.048	2.467	2.763	3.408	3.674
29	1.311	1.699	2.045	2.462	2.756	3.396	3.660
30	1.310	1.697	2.042	2.457	2.750	3.385	3.646
60	1.296	1.671	2.000	2.390	2.660	3.232	3.460
120	1.289	1.658	1.980	2.358	2.617	3.160	3.373
∞	1.282	1.645	1.960	2.326	2.576	3.091	3.291

APPENDIX 2

F DISTRIBUTION

F Distribution critical values for P=0.10

↓ Denominator												
	Numerator DF											
DF	**1**	**2**	**3**	**4**	**5**	**7**	**10**	**15**	**20**	**30**	**60**	**120**
1	39.864	49.500	53.593	55.833	57.240	58.906	60.195	61.220	61.740	62.265	62.794	63.061
2	8.5264	8.9999	9.1618	9.2434	9.2926	9.3491	9.3915	9.4248	9.4413	9.4580	9.4745	9.4829
3	5.5384	5.4624	5.3907	5.3426	5.3092	5.2661	5.2304	5.2003	5.1845	5.1681	5.1513	5.1425
4	4.5448	4.3245	4.1909	4.1073	4.0505	3.9790	3.9198	3.8704	3.8443	3.8175	3.7896	3.7753
5	4.0605	3.7798	3.6194	3.5202	3.4530	3.3679	3.2974	3.2379	3.2067	3.1740	3.1402	3.1228
7	3.5895	3.2575	3.0740	2.9605	2.8833	2.7850	2.7025	2.6322	2.5947	2.5555	2.5142	2.4927
10	3.2850	2.9244	2.7277	2.6054	2.5216	2.4139	2.3226	2.2434	2.2007	2.1554	2.1071	2.0818
15	3.0731	2.6951	2.4898	2.3615	2.2729	2.1582	2.0593	1.9722	1.9243	1.8727	1.8168	1.7867
20	2.9746	2.5893	2.3801	2.2490	2.1582	2.0397	1.9368	1.8450	1.7939	1.7383	1.6768	1.6432
30	2.8808	2.4887	2.2761	2.1423	2.0493	1.9269	1.8195	1.7222	1.6674	1.6064	1.5376	1.4990
60	2.7911	2.3932	2.1774	2.0409	1.9457	1.8194	1.7070	1.6034	1.5435	1.4756	1.3953	1.3476
120	2.7478	2.3473	2.1300	1.9924	1.8959	1.7675	1.6523	1.5450	1.4821	1.4094	1.3203	1.2646
500	2.7157	2.3132	2.0947	1.9561	1.8588	1.7288	1.6115	1.5009	1.4354	1.3583	1.2600	1.1937
1000	2.7106	2.3080	2.0892	1.9505	1.8530	1.7228	1.6051	1.4941	1.4281	1.3501	1.2500	1.1813

F Distribution critical values for P=0.05

↓ Denominator														
	Numerator DF													
DF	**1**	**2**	**3**	**4**	**5**	**7**	**10**	**15**	**20**	**30**	**60**	**120**	**500**	**1000**
1	161.45	199.50	215.71	224.58	230.16	236.77	241.88	245.95	248.01	250.10	252.20	253.25	254.06	254.19
2	18.513	19.000	19.164	19.247	19.296	19.353	19.396	19.429	19.446	19.462	19.479	19.487	19.494	19.495
3	10.128	9.5522	9.2766	9.1172	9.0135	8.8867	8.7855	8.7028	8.6602	8.6165	8.5720	8.5493	8.5320	8.5292
4	7.7086	6.9443	6.5915	6.3882	6.2560	6.0942	5.9644	5.8579	5.8026	5.7458	5.6877	5.6580	5.6352	5.6317
5	6.6078	5.7862	5.4095	5.1922	5.0504	4.8759	4.7351	4.6187	4.5582	4.4958	4.4314	4.3985	4.3731	4.3691
7	5.5914	4.7375	4.3469	4.1202	3.9715	3.7871	3.6366	3.5108	3.4445	3.3758	3.3043	3.2675	3.2388	3.2344
10	4.9645	4.1028	3.7082	3.4780	3.3259	3.1354	2.9782	2.8450	2.7741	2.6996	2.6210	2.5801	2.5482	2.5430
15	4.5431	3.6823	3.2874	3.0556	2.9013	2.7066	2.5437	2.4035	2.3275	2.2467	2.1601	2.1141	2.0776	2.0718
20	4.3512	3.4928	3.0983	2.8660	2.7109	2.5140	2.3479	2.2032	2.1241	2.0391	1.9463	1.8962	1.8563	1.8498
30	4.1709	3.3159	2.9223	2.6896	2.5336	2.3343	2.1646	2.0149	1.9317	1.8408	1.7396	1.6835	1.6376	1.6300
60	4.0012	3.1505	2.7581	2.5252	2.3683	2.1666	1.9927	1.8365	1.7480	1.6492	1.5343	1.4672	1.4093	1.3994
120	3.9201	3.0718	2.6802	2.4473	2.2898	2.0868	1.9104	1.7505	1.6587	1.5544	1.4289	1.3519	1.2804	1.2674
500	3.8601	3.0137	2.6227	2.3898	2.2320	2.0278	1.8496	1.6864	1.5917	1.4820	1.3455	1.2552	1.1586	1.1378
1000	3.8508	3.0047	2.6137	2.3808	2.2230	2.0187	1.8402	1.6765	1.5811	1.4705	1.3318	1.2385	1.1342	1.1096

F Distribution critical values for P=0.02

Denominator														
	Numerator DF													
DF	1	2	3	4	5	7	10	15	20	30	60	120	500	1000
1	1012.5	1249.5	1350.5	1405.8	1440.6	1481.8	1513.7	1539.1	1551.9	1564.9	1578.0	1584.6	1589.6	1590.4
2	48.505	49.000	49.166	49.249	49.299	49.356	49.398	49.432	49.448	49.465	49.482	49.490	49.496	49.497
3	20.618	18.858	18.110	17.694	17.429	17.110	16.860	16.657	16.553	16.448	16.340	16.286	16.244	16.238
4	14.040	12.142	11.344	10.899	10.616	10.274	10.003	9.7828	9.6696	9.5540	9.4359	9.3760	9.3300	9.3227
5	11.323	9.4544	8.6702	8.2330	7.9530	7.6137	7.3438	7.1234	7.0094	6.8926	6.7728	6.7119	6.6649	6.6573
7	8.9877	7.2026	6.4539	6.0347	5.7647	5.4354	5.1711	4.9531	4.8392	4.7220	4.6007	4.5384	4.4902	4.4825
10	7.6384	5.9336	5.2182	4.8157	4.5549	4.2347	3.9750	3.7580	3.6437	3.5245	3.3999	3.3354	3.2850	3.2770
15	6.7730	5.1355	4.4475	4.0584	3.8052	3.4917	3.2345	3.0168	2.9003	2.7775	2.6468	2.5780	2.5237	2.5151
20	6.3907	4.7875	4.1134	3.7312	3.4817	3.1713	2.9149	2.6955	2.5771	2.4509	2.3148	2.2421	2.1841	2.1747
30	6.0382	4.4695	3.8093	3.4339	3.1877	2.8803	2.6239	2.4020	2.2805	2.1493	2.0047	1.9255	1.8611	1.8505
60	5.7127	4.1785	3.5319	3.1633	2.9207	2.6157	2.3586	2.1326	2.0067	1.8676	1.7085	1.6169	1.5383	1.5251
120	5.5594	4.0423	3.4026	3.0372	2.7963	2.4923	2.2347	2.0059	1.8769	1.7322	1.5613	1.4577	1.3629	1.3458
500	5.4467	3.9428	3.3083	2.9453	2.7057	2.4024	2.1441	1.9128	1.7809	1.6307	1.4468	1.3273	1.2019	1.1750
1000	5.4293	3.9274	3.2937	2.9311	2.6915	2.3884	2.1300	1.8983	1.7659	1.6146	1.4284	1.3053	1.1701	1.1388

F Distribution critical values for P=0.01

Denominator														
	Numerator DF													
DF	1	2	3	4	5	7	10	15	20	30	60	120	500	1000
1	4052.2	4999.5	5403.4	5624.6	5763.6	5928.4	6055.8	6157.3	6208.7	6260.6	6313.0	6339.4	6359.5	6362.7
2	98.503	99.000	99.166	99.249	99.299	99.356	99.399	99.433	99.449	99.466	99.482	99.491	99.497	99.498
3	34.116	30.817	29.457	28.710	28.237	27.672	27.229	26.872	26.690	26.504	26.316	26.221	26.148	26.137
4	21.198	18.000	16.694	15.977	15.522	14.976	14.546	14.198	14.020	13.838	13.652	13.558	13.486	13.474
5	16.258	13.274	12.060	11.392	10.967	10.455	10.051	9.7222	9.5526	9.3793	9.2020	9.1118	9.0424	9.0314
7	12.246	9.5467	8.4513	7.8466	7.4605	6.9929	6.6201	6.3143	6.1554	5.9920	5.8236	5.7373	5.6707	5.6601
10	10.044	7.5594	6.5523	5.9944	5.6363	5.2001	4.8492	4.5582	4.4055	4.2469	4.0818	3.9964	3.9303	3.9195
15	8.6831	6.3588	5.4169	4.8932	4.5557	4.1416	3.8049	3.5223	3.3719	3.2141	3.0471	2.9594	2.8906	2.8796
20	8.0960	5.8489	4.9382	4.4306	4.1027	3.6987	3.3682	3.0880	2.9377	2.7785	2.6078	2.5167	2.4446	2.4330
30	7.5624	5.3903	4.5098	4.0179	3.6990	3.3046	2.9791	2.7002	2.5486	2.3859	2.2078	2.1108	2.0321	2.0192
60	7.0771	4.9774	4.1259	3.6491	3.3388	2.9530	2.6318	2.3522	2.1978	2.0284	1.8362	1.7264	1.6328	1.6169
120	6.8509	4.7865	3.9490	3.4795	3.1736	2.7918	2.4720	2.1914	2.0345	1.8600	1.6557	1.5330	1.4215	1.4015
500	6.6858	4.6479	3.8210	3.3569	3.0539	2.6751	2.3564	2.0746	1.9152	1.7353	1.5175	1.3774	1.2317	1.2007
1000	6.6603	4.6264	3.8012	3.3379	3.0356	2.6571	2.3387	2.0564	1.8967	1.7158	1.4953	1.3513	1.1947	1.1586

F Distribution critical values for P=0.005

Denominator														
	Numerator DF													
DF	**1**	**2**	**3**	**4**	**5**	**7**	**10**	**15**	**20**	**30**	**60**	**120**	**500**	**1000**
1	16211	19999	21615	22500	23056	23715	24224	24630	24836	25044	25253	25359	25439	25452
2	198.50	199.00	199.17	199.25	199.30	199.36	199.40	199.43	199.45	199.47	199.48	199.49	199.50	199.50
3	55.552	49.799	47.467	46.195	45.392	44.434	43.686	43.085	42.777	42.466	42.149	41.989	41.867	41.848
4	31.333	26.284	24.259	23.155	22.456	21.622	20.967	20.438	20.167	19.891	19.611	19.468	19.359	19.342
5	22.785	18.314	16.530	15.556	14.940	14.200	13.618	13.146	12.903	12.656	12.402	12.274	12.175	12.159
7	16.235	12.404	10.882	10.050	9.5221	8.8853	8.3803	7.9677	7.7539	7.5345	7.3088	7.1933	7.1044	7.0902
10	12.826	9.4270	8.0807	7.3428	6.8723	6.3025	5.8467	5.4706	5.2740	5.0705	4.8592	4.7501	4.6656	4.6521
15	10.798	7.7007	6.4760	5.8029	5.3721	4.8473	4.4235	4.0697	3.8826	3.6868	3.4802	3.3722	3.2874	3.2739
20	9.9439	6.9865	5.8176	5.1744	4.7616	4.2569	3.8470	3.5020	3.3178	3.1234	2.9159	2.8058	2.7186	2.7046
30	9.1796	6.3547	5.2387	4.6233	4.2275	3.7416	3.3439	3.0058	2.8231	2.6277	2.4151	2.2998	2.2066	2.1914
60	8.4946	5.7950	4.7290	4.1399	3.7599	3.2911	2.9042	2.5705	2.3872	2.1874	1.9621	1.8341	1.7256	1.7073
120	8.1789	5.5393	4.4972	3.9207	3.5482	3.0874	2.7052	2.3728	2.1882	1.9840	1.7468	1.6055	1.4778	1.4550
500	7.9498	5.3548	4.3304	3.7632	3.3963	2.9414	2.5625	2.2303	2.0441	1.8352	1.5844	1.4245	1.2595	1.2247
1000	7.9145	5.3265	4.3049	3.7391	3.3731	2.9191	2.5406	2.2084	2.0219	1.8121	1.5584	1.3945	1.2176	1.1771

F Distribution critical values for P=0.001

Denominator														
	Numerator DF													
DF	**1**	**2**	**3**	**4**	**5**	**7**	**10**	**15**	**20**	**30**	**60**	**120**	**500**	**1000**
1	405284	499999	540379	562500	576405	592873	605621	615764	620908	626099	631337	633972	635983	63630
2	998.50	999.00	999.17	999.25	999.30	999.36	999.40	999.43	999.45	999.47	999.48	999.49	999.50	999.5
3	167.03	148.50	141.11	137.10	134.58	131.58	129.25	127.37	126.42	125.45	124.47	123.97	123.59	123.5
4	74.137	61.245	56.177	53.436	51.712	49.658	48.053	46.761	46.100	45.429	44.746	44.400	44.135	44.09
5	47.181	37.122	33.202	31.085	29.752	28.163	26.917	25.911	25.395	24.869	24.333	24.061	23.852	23.81
7	29.245	21.689	18.772	17.198	16.206	15.019	14.083	13.324	12.932	12.530	12.119	11.909	11.747	11.72
10	21.040	14.905	12.553	11.283	10.481	9.5174	8.7539	8.1288	7.8038	7.4688	7.1224	6.9443	6.8065	6.784
15	16.587	11.339	9.3352	8.2526	7.5673	6.7408	6.0808	5.5351	5.2484	4.9502	4.6378	4.4749	4.3478	4.327
20	14.819	9.9526	8.0984	7.0960	6.4606	5.6920	5.0753	4.5618	4.2900	4.0051	3.7030	3.5439	3.4184	3.398
30	13.293	8.7734	7.0544	6.1245	5.5339	4.8173	4.2389	3.7528	3.4928	3.2171	2.9197	2.7595	2.6310	2.610
60	11.973	7.7678	6.1712	5.3067	4.7565	4.0864	3.5415	3.0781	2.8265	2.5549	2.2522	2.0821	1.9390	1.915
120	11.380	7.3212	5.7814	4.9471	4.4157	3.7669	3.2372	2.7833	2.5345	2.2621	1.9502	1.7668	1.6027	1.573
500	10.957	7.0041	5.5056	4.6935	4.1757	3.5424	3.0234	2.5759	2.3282	2.0538	1.7292	1.5260	1.3191	1.275
1000	10.892	6.9556	5.4637	4.6549	4.1392	3.5083	2.9909	2.5443	2.2968	2.0217	1.6946	1.4866	1.2661	1.216

APPENDIX 3

TUKEY HONESTLY SIGNIFICANT DIFFERENCE TABLE

The studentized range statistic (q)*

*The critical values for q corresponding to alpha = .05 (top) and alpha = .01 (bottom)

df for Error Term	k= Number of Treatments								
	2	3	4	5	6	7	3	9	10
5	3.64 5.70	4.60 6.98	5.22 7.80	5.67 8.42	6.03 8.91	6.33 9.32	6.58 9.67	6.80 9.97	6.99 10.24
6	3.46 5.24	4.34 6.33	4.90 7.03	5.30 7.56	5.63 7.97	5.90 8.32	6.12 8.61	6.32 8.87	6.49 9.10
7	3.34 4.95	4.16 5.92	4.68 6.54	5.06 7.01	5.36 7.37	5.61 7.68	5.82 7.94	6.00 8.17	6.16 8.37
8	3.26 4.75	4.04 5.64	4.53 6.20	4.89 6.62	5.17 6.96	5.40 7.24	5.60 7.47	5.77 7.68	5.92 7.86
9	3.20 4.60	3.95 5.43	4.41 5.96	4.76 6.35	5.02 6.66	5.24 6.91	5.43 7.13	5.59 7.33	5.74 7.49
10	3.15 4.48	3.88 5.27	4.33 5.77	4.65 6.14	4.91 6.43	5.12 6.67	5.30 6.87	5.46 7.05	5.60 7.21
11	3.11 4.39	3.82 5.15	4.26 5.62	4.57 5.97	4.82 6.25	5.03 6.48	5.20 6.67	5.35 6.84	5.49 6.99
12	3.08 4.32	3.77 5.05	4.20 5.50	4.51 5.84	4.75 6.10	4.95 6.32	5.12 6.51	5.27 6.67	5.39 6.81
13	3.06 4.26	3.73 4.96	4.15 5.40	4.45 5.73	4.69 5.98	4.88 6.19	5.05 6.37	5.19 6.53	5.32 6.67
14	3.03 4.21	3.70 4.89	4.11 5.32	4.41 5.63	4.64 5.88	4.83 6.08	4.99 6.26	5.13 6.41	5.25 6.54
15	3.01 4.17	3.67 4.84	4.08 5.25	4.37 5.56	4.59 5.80	4.78 5.99	4.94 6.16	5.08 6.31	5.20 6.44
16	3.00 4.13	3.65 4.79	4.05 5.19	4.33 5.49	4.56 5.72	4.74 5.92	4.90 6.08	5.03 6.22	5.15 6.35
17	2.98 4.10	3.63 4.74	4.02 5.14	4.30 5.43	4.52 5.66	4.70 5.85	4.86 6.01	4.99 6.15	5.11 6.27
18	2.97 4.07	3.61 4.70	4.00 5.09	4.28 5.38	4.49 5.60	4.67 5.79	4.82 5.94	4.96 6.08	5.07 6.20

19	2.96 4.05	3.59 4.67	3.98 5.05	4.25 5.33	4.47 5.55	4.65 5.73	4.79 5.89	4.92 6.02	5.04 6.14
20	2.95 4.02	3.58 4.64	3.96 5.02	4.23 5.29	4.45 5.51	4.62 5.69	4.77 5.84	4.90 5.97	5.01 6.09
24	2.92 3.96	3.53 4.55	3.90 4.91	4.17 5.17	4.37 5.37	4.54 5.54	4.68 5.69	4.81 5.81	4.92 5.92
30	2.89 3.89	3.49 4.45	3.85 4.80	4.10 5.05	4.30 5.24	4.46 5.40	4.60 5.54	4.72 5.65	4.82 5.76
40	2.86 3.82	3.44 4.37	3.79 4.70	4.04 4.93	4.23 5.11	4.39 5.26	4.52 5.39	4.63 5.50	4.73 5.60
60	2.83 3.76	3.40 4.28	3.74 4.59	3.98 4.82	4.16 4.99	4.31 5.13	4.44 5.25	4.55 5.36	4.65 5.45
120	2.80 3.70	3.36 4.20	3.68 4.50	3.92 4.71	4.10 4.87	4.24 5.01	4.36 5.12	4.47 5.21	4.56 5.30
infinity	2.77 3.64	3.31 4.12	3.63 4.40	3.86 4.60	4.03 4.76	4.17 4.88	4.29 4.99	4.39 5.08	4.47 5.16

APPENDIX 4

NORMAL DISTRIBUTION TABLE—Z PROCEDURES

Standard Normal Distribution

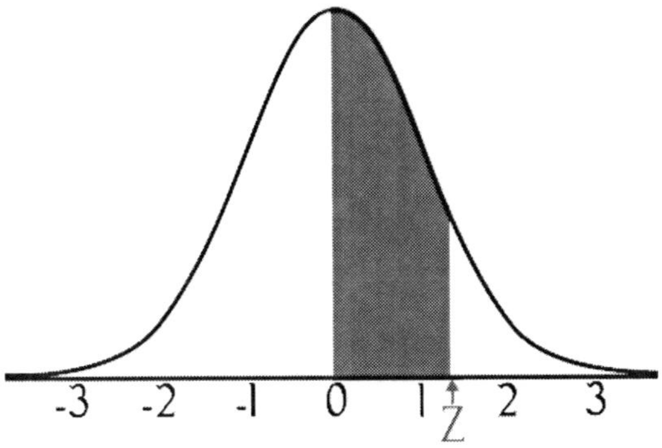

This is the "bell-shaped" curve of the Standard Normal Distrubution.

The table below can be used to find the area under the curve from the central line to any "Z-score" value up to 3, in steps of 0.1

This will then tell you what portion of the population are within "Z" standard deviations of the mean.

Instead of one LONG table, we have put the 0.1 accuracies running down, then the 0.01 accuracies running along.

For example, to determine the area under the curve between 0 and 0.45, start at the row for 0.4, and read along until 0.45 - there is the value 0.1736

Because the curve is symmetrical, the same table can be used for values going either direction, so a negative 0.45 also has an area of 0.1736

	0.00	**0.01**	**0.02**	**0.03**	**0.04**	**0.05**	**0.06**	**0.07**	**0.08**	**0.09**
0.0	0.0000	0.0040	0.0080	0.0120	0.0160	0.0199	0.0239	0.0279	0.0319	0.0359
0.1	0.0398	0.0438	0.0478	0.0517	0.0557	0.0596	0.0636	0.0675	0.0714	0.0753
0.2	0.0793	0.0832	0.0871	0.0910	0.0948	0.0987	0.1026	0.1064	0.1103	0.1141
0.3	0.1179	0.1217	0.1255	0.1293	0.1331	0.1368	0.1406	0.1443	0.1480	0.1517
0.4	0.1554	0.1591	0.1628	0.1664	0.1700	0.1736	0.1772	0.1808	0.1844	0.1879
0.5	0.1915	0.1950	0.1985	0.2019	0.2054	0.2088	0.2123	0.2157	0.2190	0.2224
0.6	0.2257	0.2291	0.2324	0.2357	0.2389	0.2422	0.2454	0.2486	0.2517	0.2549
0.7	0.2580	0.2611	0.2642	0.2673	0.2704	0.2734	0.2764	0.2794	0.2823	0.2852
0.8	0.2881	0.2910	0.2939	0.2967	0.2995	0.3023	0.3051	0.3078	0.3106	0.3133
0.9	0.3159	0.3186	0.3212	0.3238	0.3264	0.3289	0.3315	0.3340	0.3365	0.3389
1.0	0.3413	0.3438	0.3461	0.3485	0.3508	0.3531	0.3554	0.3577	0.3599	0.3621
1.1	0.3643	0.3665	0.3686	0.3708	0.3729	0.3749	0.3770	0.3790	0.3810	0.3830

1.2	0.3849	0.3869	0.3888	0.3907	0.3925	0.3944	0.3962	0.3980	0.3997	0.4015
1.3	0.4032	0.4049	0.4066	0.4082	0.4099	0.4115	0.4131	0.4147	0.4162	0.4177
1.4	0.4192	0.4207	0.4222	0.4236	0.4251	0.4265	0.4279	0.4292	0.4306	0.4319
1.5	0.4332	0.4345	0.4357	0.4370	0.4382	0.4394	0.4406	0.4418	0.4429	0.4441
1.6	0.4452	0.4463	0.4474	0.4484	0.4495	0.4505	0.4515	0.4525	0.4535	0.4545
1.7	0.4554	0.4564	0.4573	0.4582	0.4591	0.4599	0.4608	0.4616	0.4625	0.4633
1.8	0.4641	0.4649	0.4656	0.4664	0.4671	0.4678	0.4686	0.4693	0.4699	0.4706
1.9	0.4713	0.4719	0.4726	0.4732	0.4738	0.4744	0.4750	0.4756	0.4761	0.4767
2.0	0.4772	0.4778	0.4783	0.4788	0.4793	0.4798	0.4803	0.4808	0.4812	0.4817
2.1	0.4821	0.4826	0.4830	0.4834	0.4838	0.4842	0.4846	0.4850	0.4854	0.4857
2.2	0.4861	0.4864	0.4868	0.4871	0.4875	0.4878	0.4881	0.4884	0.4887	0.4890
2.3	0.4893	0.4896	0.4898	0.4901	0.4904	0.4906	0.4909	0.4911	0.4913	0.4916
2.4	0.4918	0.4920	0.4922	0.4925	0.4927	0.4929	0.4931	0.4932	0.4934	0.4936
2.5	0.4938	0.4940	0.4941	0.4943	0.4945	0.4946	0.4948	0.4949	0.4951	0.4952
2.6	0.4953	0.4955	0.4956	0.4957	0.4959	0.4960	0.4961	0.4962	0.4963	0.4964
2.7	0.4965	0.4966	0.4967	0.4968	0.4969	0.4970	0.4971	0.4972	0.4973	0.4974
2.8	0.4974	0.4975	0.4976	0.4977	0.4977	0.4978	0.4979	0.4979	0.4980	0.4981
2.9	0.4981	0.4982	0.4982	0.4983	0.4984	0.4984	0.4985	0.4985	0.4986	0.4986
3.0	0.4987	0.4987	0.4987	0.4988	0.4988	0.4989	0.4989	0.4989	0.4990	0.4990

APPENDIX 5

CHI-SQUARE DISTRIBUTION

Chi-square distribution table

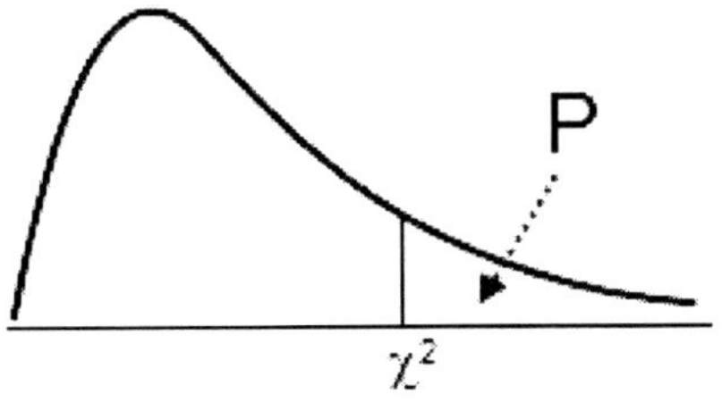

	P										
DF	**0.995**	**0.975**	**0.20**	**0.10**	**0.05**	**0.025**	**0.02**	**0.01**	**0.005**	**0.002**	**0.001**
1	0.0000393	0.000982	1.642	2.706	3.841	5.024	5.412	6.635	7.879	9.550	10.828
2	0.0100	0.0506	3.219	4.605	5.991	7.378	7.824	9.210	10.597	12.429	13.816
3	0.0717	0.216	4.642	6.251	7.815	9.348	9.837	11.345	12.838	14.796	16.266
4	0.207	0.484	5.989	7.779	9.488	11.143	11.668	13.277	14.860	16.924	18.467
5	0.412	0.831	7.289	9.236	11.070	12.833	13.388	15.086	16.750	18.907	20.515
6	0.676	1.237	8.558	10.645	12.592	14.449	15.033	16.812	18.548	20.791	22.458
7	0.989	1.690	9.803	12.017	14.067	16.013	16.622	18.475	20.278	22.601	24.322
8	1.344	2.180	11.030	13.362	15.507	17.535	18.168	20.090	21.955	24.352	26.124
9	1.735	2.700	12.242	14.684	16.919	19.023	19.679	21.666	23.589	26.056	27.877
10	2.156	3.247	13.442	15.987	18.307	20.483	21.161	23.209	25.188	27.722	29.588
11	2.603	3.816	14.631	17.275	19.675	21.920	22.618	24.725	26.757	29.354	31.264
12	3.074	4.404	15.812	18.549	21.026	23.337	24.054	26.217	28.300	30.957	32.909
13	3.565	5.009	16.985	19.812	22.362	24.736	25.472	27.688	29.819	32.535	34.528
14	4.075	5.629	18.151	21.064	23.685	26.119	26.873	29.141	31.319	34.091	36.123
15	4.601	6.262	19.311	22.307	24.996	27.488	28.259	30.578	32.801	35.628	37.697
16	5.142	6.908	20.465	23.542	26.296	28.845	29.633	32.000	34.267	37.146	39.252
17	5.697	7.564	21.615	24.769	27.587	30.191	30.995	33.409	35.718	38.648	40.790
18	6.265	8.231	22.760	25.989	28.869	31.526	32.346	34.805	37.156	40.136	42.312
19	6.844	8.907	23.900	27.204	30.144	32.852	33.687	36.191	38.582	41.610	43.820
20	7.434	9.591	25.038	28.412	31.410	34.170	35.020	37.566	39.997	43.072	45.315
21	8.034	10.283	26.171	29.615	32.671	35.479	36.343	38.932	41.401	44.522	46.797
22	8.643	10.982	27.301	30.813	33.924	36.781	37.659	40.289	42.796	45.962	48.268
23	9.260	11.689	28.429	32.007	35.172	38.076	38.968	41.638	44.181	47.391	49.728
24	9.886	12.401	29.553	33.196	36.415	39.364	40.270	42.980	45.559	48.812	51.179
25	10.520	13.120	30.675	34.382	37.652	40.646	41.566	44.314	46.928	50.223	52.620
26	11.160	13.844	31.795	35.563	38.885	41.923	42.856	45.642	48.290	51.627	54.052
27	11.808	14.573	32.912	36.741	40.113	43.195	44.140	46.963	49.645	53.023	55.476
28	12.461	15.308	34.027	37.916	41.337	44.461	45.419	48.278	50.993	54.411	56.892
29	13.121	16.047	35.139	39.087	42.557	45.722	46.693	49.588	52.336	55.792	58.301
30	13.787	16.791	36.250	40.256	43.773	46.979	47.962	50.892	53.672	57.167	59.703

APPENDIX 6

MANN-WHITNEY U-TEST

A table of Mann-Whitney critical *U* values is given below for alpha=0.05 (1-tailed=directional H_a) and selected sample sizes only.

$n_i \backslash n_j$	**2**	**3**	**4**	**5**	**6**	**7**	**8**	**9**	**10**	**15**	**20**
2	0	0	0	1	1	1	2	2	2	4	5
3	0	1	1	2	3	3	4	5	5	8	12
4	0	1	2	3	4	5	6	7	8	13	19
5	1	2	3	5	6	7	9	10	12	19	26
6	1	3	4	6	8	9	11	13	15	24	33
7	1	3	5	7	9	12	14	16	18	29	40
8	2	4	6	9	11	14	16	19	21	34	48
9	2	5	7	10	13	16	19	22	25	40	55
10	2	5	8	12	15	18	21	25	28	45	63
15	4	8	13	19	24	29	34	40	45	73	101
20	5	12	19	26	33	40	48	55	63	101	139

APPENDIX 7

WILCOXON MATCHED-PAIRS

SIGNED RANKS t-TEST

Selected Critical Values of the Wilcoxon Statistic (*W*) for alpha = .1 and .05		
Number of Pairs	Alpha Level	
	.1	.05
5	0	-
6	2	0
7	3	2
8	5	3
9	8	5
10	10	8
12	17	13
15	30	25
20	60	52
25	100	89
Note: W is significant at the chosen alpha level if it is **smaller** than the critical value given in the table.		

APPENDIX 8

SPEARMAN r_s

If the obtained *rs* is equal to or larger than the value shown, then its *p* level is equal to or smaller than the level shown at the top of the column (.05 or .01).

	Probability level	
Number of pairs (*N*)	**.05**	**.01**
7	**.714**	**.893**
8	**.643**	**.833**
9	**.600**	**.783**
10	**.564**	**.746**
12	**.506**	**.712**
14	**.456**	**.645**
16	**.425**	**.601**
18	**.399**	**.564**
20	**.377**	**.534**
22	**.359**	**.508**
24	**.343**	**.485**
26	**.329**	**.465**
28	**.317**	**.448**
30	**.306**	**.432**

APPENDIX 9

DISTRIBUTION USED TO DETERMINE CRITICAL VALUE

	Normal Distribution	t Distribution	F Distribution	Chi-Square Distribution	**
Independent t-Test		XX	(XX)		
One Way ANOVA			XX		
Paired t-Test		XX			
Pearson Correlation		XX	(XX)		XX
Mann-Whitney U-Test	XX				XX
Kruskal Wallis Test				XX	
Wilcoxon Signed Ranks Test	XX				
Spearman Rank Correlation		(XX)	XX		XX
Friedman Two-Way ANOVA			(XX)	XX	
Chi-Square Goodness of Fit				XX	
Chi-Square Test Of Independence				XX	
Tukey HSD Test					XX

APPENDIX 10

SAMPLE SIZE DATA

Sample Sizes (S) Required for Given Population Sizes (N)

N	S	N	S	N	S	N	S	N	S
10	10	100	80	280	162	800	260	2800	338
15	14	110	86	290	165	850	265	3000	341
20	19	120	92	300	169	900	269	3500	346
25	24	130	97	320	175	950	274	4000	351
30	28	140	103	340	181	1000	278	4500	354
35	32	150	108	360	186	1100	285	5000	357
40	36	160	113	380	191	1200	291	6000	361
45	40	170	118	400	196	1300	297	7000	364
50	44	180	123	420	201	1400	302	8000	367
55	48	190	127	440	205	1500	306	9000	368
60	52	200	132	460	210	1600	310	10000	370
65	56	210	136	480	214	1700	313	15000	375
70	59	220	140	500	217	1800	317	20000	377
75	63	230	144	550	226	1900	320	30000	379
80	66	240	148	600	234	2000	322	40000	380
85	70	250	152	650	242	2200	327	50000	381
90	73	260	155	700	248	2400	331	75000	382
95	76	270	159	750	254	2600	335	100000	384

Note. From r. V. Krejcie and D. W. Morgan, Determining sample size for research activities. *Educational and Psychological Measurement,* 30, 608. Copyright © 1970 by Sage Publications.

Sample size formula

$$n = \left[\chi^2 N\Phi(1-\Phi)\right] / \left[d^2(N-1) + \chi^2\Phi(1-\Phi)\right]$$

χ^2 = the table value of Chi-square for one degree of freedom and the desired confidence interval – (usually 90 or 95 percent) – (for 95 percent [0.05] use3.841)

N = the population size or sample size of the smallest sub-group to be proportionately represented.

φ = the population proportion which it is desired to estimate (.50 will give minimum sample size).

d = the degree of accuracy expressed as a proportion (.05 with confidence interval of 90 percent; .025 with confidence interval of 95 percent.)

APPENDIX 11

RANDOM NUMBERS TABLE

How To Use A Random Number Table

1. Number each member of the population.
2. Determine sample size; this is the number of random numbers you will select
3. Determine starting point in table by randomly picking a page and dropping your finger on the page with your eyes closed.
4. Choose a direction in which to read (up to down, left to right, or right to left).
5. Select the first number; read from the table whose last digits are between 0 and your sample size number. (If sample size is a two digit number, then your number would be the last 2 digits; if it is a four digit number, your number would be the last 4 digits; etc.).
6. Once a number is chosen, do not use it again.
7. If you reach the end of the table before obtaining the number of random numbers in your sample, pick another starting point, read in a different direction and continue until done.

Example: Sample Size = 50; starting point is column 3, row 2 on Random Number Table (next page); read down. You would select population numbers 4, 7, 30, 14, etc., until you had 50 unique numbers.

03 04

38 67

69 84

27 30

36 14

etc...

20 17	42 28	23 17	59 66	38	61	02 10	86 10	51 55	92 52	44 25
74 49	04 19	03 04	10 33	53	70	11 54	48 63	94 60	94 49	57 38

94 70	49 31	38 67	23 42	29	65	40 88	78 71	37 18	48 64	06 57
22 15	78 15	69 84	32 52	32	54	15 12	54 02	01 37	38 37	12 93
93 29	12 18	27 30	30 55	91	87	50 57	58 51	49 36	12 53	96 40
45 04	77 97	36 14	99 45	52	95	69 85	03 83	51 87	85 56	22 37
44 91	99 49	89 39	94 60	48	49	06 77	64 72	59 26	08 51	25 57
16 23	91 02	19 96	47 59	89	65	27 84	30 92	63 37	26 24	23 66
04 50	65 04	65 65	82 42	70	51	55 04	61 47	88 83	99 34	82 37
32 70	17 72	03 61	66 26	24	71	22 77	88 33	17 78	08 92	73 49
03 64	59 07	42 95	81 39	06	41	20 81	92 34	51 90	39 08	21 42
62 49	00 90	67 86	93 48	31	83	19 07	67 68	49 03	27 47	52 03
61 00	95 86	98 36	14 03	48	88	51 07	33 40	06 86	33 76	68 57
89 03	90 49	28 74	21 04	09	96	60 45	22 03	52 80	01 79	33 81
01 72	33 85	52 40	60 07	06	71	89 27	14 29	55 24	85 79	31 96
27 56	49 79	34 34	32 22	60	53	91 17	33 26	44 70	93 14	99 70
49 05	74 48	10 55	35 25	24	28	20 22	35 66	66 34	26 35	91 23
49 74	37 25	97 26	33 94	42	23	01 28	59 58	92 69	03 66	73 82
20 26	22 43	88 08	19 85	08	12	47 65	65 63	56 07	97 85	56 79
48 87	77 96	43 39	76 93	08	79	22 18	54 55	93 75	97 26	90 77
08 72	87 46	75 73	00 11	27	07	05 20	30 85	22 21	04 67	19 13
95 97	98 62	17 27	31 42	64	71	46 22	32 75	19 32	20 99	94 85
37 99	57 31	70 40	46 55	46	12	24 32	36 74	69 20	72 10	95 93
05 79	58 37	85 33	75 18	88	71	23 44	54 28	00 48	96 23	66 45

APPENDIX 12

NETLIBRARY BIBLIOGRAPHY FOR STATISTICS

Updated October, 2009

Altman, M. (n.d.). *Numerical issues in statistical computing for the social scientist.* Hoboken, NJ: Wiley.

Johnson, V. E. (1999). *Ordinal data modeling.* NY: Springer

Levin, I. P. (1999). *Relating statistics and experimental design: an introduction.* Thousand Oaks, CA: Sage Publications.

Maier, M. (1950). *The data game: controversies in social science statistics.* Armonk: NY: M.E. Sharpe.

Shao, J. (1999). *Mathematical statistics.* NY: Springer.

Shorack, Galen R. (2000). *Probability for statisticians.* NY: Springer.

Spirer, H. F. (1998). *Misused statistics.* NY: M. Dekker.

Stevens, J. (1999). *Intermediate statistics: a modern approach.* Mahwah, NJ: Lawrence Erlbaum Associates.

Ullah, A. (Ed.). (1998). *Handbook of applied economic statistics.* New York: Marcel Dekker.

APPENDIX 13

WORKS CITED

Bulmer, M. G. (1979). *Principles of statistics.* New York: Dover Publications, Inc.

Elmore, P. B. & Woehlke, P. L. (1997). *Basic statistics.* New York: Longman, An imprint of Addison Wesley Longman, Inc.

Field, M. H. (1996). *Statistics for science projects.* Martin, Tennessee: Center for Environmental and Conservation Education. Unpublished.

Guilford, J.P. (1965). *Fundamental statistics in psychology and education.* St. Louis: McGraw-Hill Book Company.

Halvorson, M. & Young, M. (1997). *Running microsoft office 97.* Redmond, Washington: Microsoft Press.

Hinton, P. (1996). *Statistics explained: a guide for social science students.* New York: Routledge.

Kitchens, L. (1987). *Exploring statistics: a modern introduction to data analysis and Inference.* New York: West Publishing Company.

Levin, J. (1983). *Elementary statistics in social research.* New York: Harper and Row, Publishers.

Siegel, S. (1956). *Nonparametric Statistics for the behavioral sciences.* New York: McGraw-Hill Book Company.

Tai, S. W. (1978). *Social science statistics: its elements and applications.* Santa Monica, California: Goodyear Publishing Company, Inc.

Yaremko, R.M., Harari, H., Harrison, r. C., & Lynn, E. (1982). *Reference handbook of research and statistical methods in psychology: for students and professions.*

$69.95
ISBN 978-1-61539-808-9
56995>
9 781615 398089

Breinigsville, PA USA
23 November 2010
249942BV00001B/14/P